Letters from an American Farmer

Letters from an American Farmer

The Eastern European and Russian Correspondence of Roswell Garst

Edited by Richard Lowitt
and Harold Lee

Northern Illinois University Press

DeKalb, Illinois 1987

Library of Congress
Cataloging-in-Publication Data
Garst, Roswell, 1898?–1977.
Letters from an American farmer.
Bibliography: p.
Includes index
1. Garst, Roswell, 1898?–1977.
2. Agriculture—Europe, Eastern.
3. Agriculture—Soviet Union.
4. Agriculture—United States.
5. Agriculturists—Iowa—
Coon Rapids—Correspondence.
6. Coon Rapids (Iowa)—Biography.
I. Lowitt, Richard, 1922–
II. Lee, Harold. III. Title.
S417.G26A4
1987   630'.947   86-21698
ISBN 0-87580-123-4

To John Chrystal
who carries on the tradition

Contents

USSR, 1969–77

Introduction

Roswell Garst achieved sudden and spectacular international recognition when Nikita Khrushchev chose to visit the Garst farm near Coon Rapids, Iowa, during the Soviet Premier's American tour in the fall of 1959. Millions of television viewers watched the leather-faced, strong-jawed, and embattled Iowa agriculturalist conduct a bemused Khrushchev through swarming hordes of journalists to show him the latest techniques in American cattle feeding. United States Ambassador to the United Nations Henry Cabot Lodge, official host for the visit, improvised a loudspeaker enabling the enveloping crowd to hear Garst describe the grinding mill for corncobs, the highly mechanized process for feeding this protein-enriched cellulose to livestock in nearby feedlots, and the quality of the corn and cattle they saw. They also heard Garst and Khrushchev, both already acquainted and both veteran showmen, trade wit and repartee, and, at a nearby farm, witnessed Garst's anger at the obstructive mass of newsmen explode in an assault on them with handfuls of silage that produced the most memorable image of the day, and possibly of Khrushchev's entire American visit.

Despite its ambiance of what a later generation would describe as a "media event," the Soviet premier's afternoon down on the farm in Coon Rapids reflects the stature Garst had attained as "one of the most important figures in the establishment of East-West relations," as Silviu Brucan, a former Rumanian diplomat, has noted. The distinguished journalist Harrison E. Salisbury has written that Garst, "though not a diplomat, became a remarkable instrument of diplomacy." In Salisbury's view, Garst played a role on the international stage "almost unique for his time."[1]

The story of this phase of Garst's long career—he was sixty-one years

1. Harold Lee, *Roswell Garst: A Biography* (Ames, Iowa State University Press, 1984), pp. ix, xiv.

old when Khrushchev came to Coon Rapids—began four years earlier. In January 1955, Khrushchev, First Secretary of the Central Committee of the Communist Party of the Soviet Union, remarked in a major address that what the Soviet Union needed was an Iowa corn belt. Lauren Soth, in a Pulitzer Prize-winning editorial that appeared in the *Des Moines Register* for 10 February 1955, "without diplomatic authority of any kind" extended an invitation "to any delegation Khrushchev wants to select to come to Iowa" and promised that "everything we Iowans know about corn, other feed grains, forage crops, meat animals, and the dairy and poultry industries will be available to the Russians for the asking."

To the apparent dismay of the State Department, Khrushchev accepted the invitation offered in Soth's editorial. The State Department insisted upon an exchange of delegations and again Khrushchev accepted. From this official arrangement came the unofficial visit to the Garst farm, Garst's subsequent invitation to visit the Soviet Union, and the beginning of Garst's extensive correspondence pertaining almost exclusively to agricultural matters with Eastern European and Russian officials. Garst's letters relate as much about developments in American agriculture as they do about conditions in Eastern Europe. Writing from Coon Rapids, Iowa, benefiting from the enhanced productivity of a vast agricultural revolution, in a nation whose officials were grappling, unsuccessfully, with the dilemma of a massive agricultural surplus but because of the cold war were unwilling to eliminate restrictions and actively encourage expanding commercial relations with the communist countries in Eastern Europe, Garst took up the challenge to become a salesman for himself, for American agriculture, and for the promotion of East-West contacts. He understood his own country's agricultural problem and also realized the necessity to help improve agriculture in war-devastated countries where productivity was not influenced by the production methods and techniques practiced in Iowa, where the dilemma of a surplus was unknown, where high-level political decisions determined agricultural responses, and where decisions made within the Kremlin affected agriculture in Kazakhstan and western Siberia. Garst continually made the point that he was not a politician, insisting that his sole objective at home and abroad was to teach, aid, and encourage people of all nations to produce more and better food with less labor.

Scholars will find in this edition of Garst's correspondence valuable source materials on such subjects as the breakthrough in East-West relations in the fifties, the transfer of agricultural technology from West to East during the ensuing thirty years, and the comparative state of Eastern European and American agricultural practices during the same period. And for those simply interested in Roswell Garst, the correspondence provides an opportunity to know him better by reading him directly and by seeing through his eyes the extraordinary events of this part of his life. The reader will have the pleasure of enjoying a lucid, entertaining, and persuasive correspondent, one of the best of his time. Garst wrote as he talked, in a manner, as John Dos Passos once described his speech, "between that of a lecturer explaining the solution of a problem at a blackboard, and a lawyer pleading with a jury." His correspondence was also voluminous and often repetitious, for he possessed an unflagging ability to make the same points over and over again. Consequently, in assembling these letters we have chosen representative sequences of his remarkable output.

Any consideration of Garst's impact in Eastern Europe must be grounded in an understanding of his life in agriculture prior to 1955. By that time, he had already been a pioneer in the introduction of hybrid seed corn in the thirties, had demonstrated the value of protein-enriched cellulose for cattle in the forties, and played a central role in the fertilizer revolution of the fifties and sixties. By 1955, Garst was a well-established businessman, innovator, educator, and leader in the new agriculture of the Midwest, and consequently superbly equipped for the role he assumed as a consequence of the first U.S.-Soviet agricultural exchange.

Garst's internationalism sprang from a complex system of strong roots, as firm and widely dispersed as those of the hybrid corn plant central to his fortunes. At the outset he was influenced by the progressive Republicanism of his uncle, Warren Garst, a devoted supporter of Albert Baird Cummins and governor of Iowa for two years after Cummins's resignation to enter the United States Senate in November 1908. Garst's own experience of the collapse of the international agricultural market after World War I and the preaching of Henry A. Wallace on the evils of the tariff wars further shaped his attitudes. Moreover, Garst's own developing educative and sales abilities, coupled with a desire to reach anyone who would listen or buy must be considered. Garst was enthusiastic about Wendell Willkie's version of "One World" and Franklin Roosevelt's fight against isolationism. Other factors were Wallace's postwar perception of the need to come to terms with the superpower confrontation that would lead to the cold war and Garst's own revulsion at the insanity of an unbridled arms race. Garst did not follow Wallace's political path, but instead chose to foster agricultural abundance for all, regardless of ideological persuasion, as the foundation of his hopes for a peaceful, fruitful international community. It was ultimately, he once remarked, just a matter of common decency.

Garst was born on 13 June 1898, the youngest son of Edward Garst, who in 1869 set up a general store in Coon Rapids, Iowa, and thereafter prospered as a merchant and as a buyer and seller of land. Garst's first—and satisfying—experience in farming was with his brother Jonathan in 1915 on a two-hundred-acre farm just south of Coon Rapids, owned but never farmed by his father, known then as Apple Farm, today as the Garst farm. After Jonathan's service in World War I, the brothers farmed new land in Canada but lost their homestead in the agricultural recession of 1920–21. Roswell returned home to turn Apple Farm into a successful dairy, and he married Elizabeth Henak in 1922. In 1926, he left the farm in the hands of a tenant and moved to Des Moines to try his hand at selling houses on a subdivision he owned jointly with Clyde B. Fletcher, a friend and mentor.

The Des Moines years (1926–30) were formative in several respects. Through Fletcher, his first entrepreneurial model outside the family, Garst discovered his talent for salesmanship. Through Henry A. Wallace, whom he met at this time, he discovered the fledgling hybrid seed corn business, which provided a focus for his continuing interest in agriculture, his penchant for salesmanship, and his curiosity. Through Wallace he also gained an understanding of the corn-hog ratio and the problems associated with protective tariffs and surpluses. By 1930 he had experienced a commercial and intellectual awakening that he would build on and bring to fruition in the coming decade.

In that year, having observed the behavior of the high-yielding, strong-stalked hybrid seed corn for four years, Garst asked Wallace for a franchise to sell the virtually unknown product in northwest Iowa, moved back to his dairy farm in Coon Rapids, raised his first fifteen acres of seed corn on it, and later sold the cows. He began his new career in the first year of the Depression, a time both daunting and exciting for Garst and his wife, Elizabeth.

The first crisis of the business was the question of raising the cash for the detasselling of the "female" rows of corn, essential in order to achieve the final cross that would produce a salable product. Elizabeth raised the money needed to pay the detassellers by offering farm holidays to city children during the summers of 1930 and 1931. Such measures could be no more than stop-gap, however, and in 1931 Garst, desperately needing operating capital and lacking assets on which to borrow, formed a partnership with Charles Thomas, a successful Coon Rapids farmer. The Garst and Thomas Hybrid Seed Corn Company was founded, survived the difficult years of the early thirties, and by 1938, under Garst's aggressive sales leadership, was beginning to dominate seed corn sales in the western corn belt. Thomas supervised the growing company's production department.

Although the growing and processing of high-quality seed were always the first priorities in the Wallace-Garst business tradition, sales were essential if the hybrid seed corn industry were to survive the Depression. Indeed, even in normal times, promoting an expensive, genetically improved seed would be a difficult task. Farmers had to be shown that it was worthwhile to buy the new seed rather than use the inferior open-pollinated kernels available in their own fields. By devising methods of demonstrating the benefits of such an investment, Garst rendered an invaluable service both to farmers and to the entire new industry, creating techniques that he would use throughout his life.

The major breakthrough for Garst and later for the Wallace organization, the Pioneer Hi-Bred Corn Company, lay in Garst's invention of the "half-the-increase" technique instituted in 1931. A farmer planted, for example, twenty rows of his own open-pollinated corn and twenty rows of hybrid corn provided by Garst or his salesmen, one type in each of the two planter boxes then used by farmers. It would be agreed that Garst and Thomas would take as payment one-half the increase in yield produced by the hybrid corn over the open-pollinated corn. Usually, the hybrid corn outperformed the open-pollinated corn by an average of at least ten bushels per acre. The farmer would thus have a graphic and profitable demonstration of the worth of the new seed corn and in fact would be charged only the cost of the corn itself rather than handing over half the increase. Garst and Thomas and Pioneer began to accumulate a steady influx of customers even during the most harrowing years of the Depression, and Garst's practice of hiring satisfied farmers as part-time salesmen spread throughout the industry.

Garst's next major contribution to the industry occurred in the late thirties, when he distributed more than ten thousand sample bags of seed corn throughout western Iowa and Nebraska, so that, as he wrote, "no man in western Iowa or eastern Nebraska can live more than three miles from a man who has had experience with it." He applied the same technique a decade later in the dissemination of fertilizer samples and again in the sixties in the distribution of planting kits for demonstration plots in Central America. In Rumania in the fifties, Garst and his sons organized farming demonstrations on several state farms. In various forms, the field demonstration was always central to Roswell Garst's teaching. In his lifelong campaign for an abundant agriculture, Garst was never an armchair general.

Garst's own burgeoning oral and epistolary gifts were aspects of his salesmanship inextricable from his technical innovations. He never ceased learning about his product and about agriculture and passing on what he learned with earthy good sense, eloquence, and enthusiasm. "I'll never forget when he came into the house," Hans Larsen, an early supervisor, recalled. "He was like a breath of fresh air. A ball of enthusiasm. He came in and expounded the philosophy of hybrid corn, what it would do for agriculture. And we forgot the severity of the Depression. The room was just filled with enthusiasm. He just had the ability to do that." By the early forties, Garst had reached his full powers as a letter writer, not only in style but also in volume and in the range of recipients, which now reached a national level. He was speaking regularly throughout the Midwest, creating, he used to say, an educated market for his product, but also beginning to consider the larger issues for agriculture raised by his personal success and the age of mechanization now firmly established in the corn belt. In addition, two of his other activities in the thirties are important themselves and essential to an understanding of Garst's later development: bringing the New Deal to the Midwestern farmer and revitalizing the tired Depression land.

Devoted to Wallace as his friend, business associate, and intellectual mentor; convinced that Wallace's approach to the agricultural crisis offered the best chance of recovery for America's farmers; anxious as any business-man to achieve prosperity once again; and bursting with newly discovered capabilities and energies, Garst threw himself during the summer and fall of 1933 into the formulation and implementation of the Agricultural Adjust-ment Administration's corn-hog program.

Concerned that Iowa farmers were not taking advantage of Secretary of Agriculture Wallace's appeal to formulate a program that would curtail corn and hog production, Garst teamed up with J. Stuart Russell of the Des Moines *Register* to organize a series of meetings within the state. Garst became chairman of the Iowa Corn-Hog Committee, then a member of the National Corn-Hog Committee, and with Wallace's personal blessing took the lead in organizing committees throughout the corn belt states. He traveled throughout that area during the summer and fall of 1933 and made several journeys to Washington to testify on behalf of producers at the meetings designed to effect an agreement between farmers and meat packers that would drastically reduce corn acreage and livestock numbers and at the same time provide adequate compensation for all parties. Throughout this period Garst remained a private citizen, not an employee of the AAA. In October a comprehensive program for corn and hogs was finally agreed upon, a program aptly described by Richard Wilson of the *Register* as "the greatest agricultural production measure yet undertaken."

In the process of making a significant contribution to this fundamental new approach to the economic, social, and political structure of the farm economy, Garst discovered his own gifts as an analyst of agriculture on a national scale, received recognition and respect in Washington, and made lifelong friends in the most liberal wing of the New Deal. Perhaps most important, he was touched deeply and permanently by that paradoxical combination of mystical and pragmatic optimism that animated New Dealers, the "new energies, new faces, young men," the "enterprising and hopeful minds" celebrated by Walter Lippmann. However, in spite of the allure of service as an appointed or elected official, Garst decided to complete the unfinished tasks awaiting him in the office of Garst and Thomas and in the black earth of Iowa farmland.

Even during his years in Des Moines, Garst had kept in touch with the land, making frequent trips to Coon Rapids to see how his tenant was doing on the two-hundred acres south of town, poking around other parts of the countryside by automobile, and beginning to sharpen a talent for spotting potential in land that became legendary. Garst's delight in a well-run farm produced profits for tenant and landlord, and invaluable experience in large-scale farming. Unable to buy farmland for himself during these years, Garst did the next best thing by managing it for others. His clients were wealthy Chicagoans and Iowans interested in owning land as a hedge against inflation and willing to put money into capital improvements for tax purposes and to ensure a reasonably good living for the tenants.

On these farms he cleared rocks, improved and expanded drainage systems, built new houses, and found profits rising steadily. Out of the experience came one significant conclusion, and a complementary and equally far-reaching consequence. Garst concluded that the more money the farmer invested in genuine improvements, the higher would be the return; this simple yet revolutionary observation replaced the widely accepted axiom that one dollar per acre was the acceptable level of investment. The consequence of this discovery was that Garst's landlords, possessed of the irrefutable evidence of increasingly large yearly checks, were willing to undertake the further and substantially higher investment required for Garst's great leap into intensive fertilization of their farmland in the early forties. "Invest, invest!" was a maxim he would pronounce again and again to Eastern Europeans in 1955.

The forties, then, were characterized by the entry of Coon Rapids into the "fertilizer era" of corn farming a decade before it became standard practice throughout the entire Midwest. Eager to find a further means of increasing yields for his own profit and to meet the need for more food he foresaw as a result of the war that had erupted in Europe in 1939, Garst plunged into the study of fertilizers, seeking information at various Mid-western universities. He discovered that nitrogen had recently been shown to be effective on corn plants when used in balance with phosphate and potash, conducted his own tests in 1942 in order to see for himself, then in 1943 launched a highly successful and profitable program on the fifty-six hundred acres he managed. Throughout the remainder of the decade Garst struggled to obtain sufficient amounts of fertilizer for himself, for the seed corn industry, and for meeting the demand he foresaw for the coming years.

During the early forties, Garst saw in the combination of hybrid corn, mechanization, and chemical fertilization (with its promise of continously-grown corn replacing crop rotation) the beginnings of an awesome increase in American agricultural production. By the end of the decade, he had participated in yet another breakthrough, the feeding of protein-enriched cellulose to cattle. Further innovations would follow, but the development of cellulose as fodder added the fourth gospel to the canon of the new Midwestern agriculture of the fifties. Hybridization, mechanization, fertil-ization, and utilization of cellulose became the fundamental texts; herbi-cides, pesticides, irrigation, and other developments would become au-thorized epistles. The evangelistic fervor with which Garst and his early associates had preached the message of corn technology was nearing its apogee.

Garst's interest in cellulose began in the backyard of Garst and Tho-mas, where a mountain of ground corncobs accumulated in the processing season during the prosperous years of the late forties. Garst learned of feeding experiments with corncobs, corn, and synthetic protein in Ohio and at the University of Wisconsin, and of research in the digestibility of cellulose begun at the Ohio Agricultural Research Station and continued at Iowa State College. Wishing to use as many of his hitherto useless corncobs as possible, Garst proved that corncobs, protein, and vitamin supplements alone provided an adequate diet for cattle. Although the use of corncobs was not taken up as widely as he had hoped, his work opened the way for the use of other forms of cellulose as feed, and his bulletins on the subject reached thousands of Midwestern farmers. By the early fifties, the Garst cattle-feeding operation was a showpiece for visitors and was much publi-cized in farm publications.

The early fifties was a time of increasing interest in international affairs for Garst. He became a champion of the Food for Peace program and pressed for its extension to eastern bloc countries. Mindful of Wallace's 1946 warnings of the need for trade as a means of easing East-West ten-sions, he actively endorsed the invitation extended in 1955 to Nikita

Khrushchev and other Soviet leaders by Lauren Soth of the Des Moines *Register* suggesting the visit of a delegation so that the Russians could find out all they wanted to know about American farming.

Although he was not a member of the delegation sent to Russia and his farm was not included in the rather hastily improvised list of those to be visited by the Russian delegation touring the United States in the summer of 1955, Garst was determined to make contact with the latter group; his opportunity arose when two of the Soviet visitors spent the night as guests of his cousin, Warren Garst, in Jefferson, Iowa. Early the next morning they made an unofficial visit to the Garst farming operation in Coon Rapids and reported what they had seen to their delegation's leader, Deputy Minister of Agriculture Vladimir Matskevich. Matskevich insisted on visiting the farm himself, was impressed by the scale of efficiency that he saw, and invited Garst to visit the Soviet Union.

Garst accepted the invitation; lectured and toured extensively in the Soviet Union, Rumania, and Hungary; and in turn entertained experts from those countries in Coon Rapids. Out of this personal exchange of delegations began a major transfer of agricultural technology to Eastern Europe. The Russians bought Pioneer seed corn to plant in spring 1956 and parent hybrid corn stock to help achieve independent production of their own hybrid and constructed hybrid seed corn plants on the Coon Rapids model. Garst also sold seed corn to Hungary, influenced the direction of their major research station, and plotted the locations for thirteen regional seed corn plants. He also sold seed to Rumania; gave impetus to the addition of a new research complex at Fundulea, near Bucharest; helped the Rumanians buy ten complete sets of farm machinery; and sent his two sons, David and Stephen, to oversee the planting and harvesting of the 1956 crop. All of those countries expanded their contacts in the United States, Canada, and Western Europe, but all considered Garst their original source of help and continued to regard him as their principal Western agricultural adviser.

It is an irony of contemporary history that the Westerner who would make the largest single contribution to the knowledge sought by Khrushchev was neither a member of the American delegation sent to Russia nor included in the list of farmers the Soviet delegation to the United States would visit. However, it is not surprising that once they had found him, the Russians, Rumanians, and Hungarians recognized immediately that here was a man whose talents and experience were of great value for their particular needs; nor is it surprising that they fell under his persuasive spell in much the same way as had Hans Larsen and the early salesmen and farmers of the thirties. For Garst, the opportunity to visit Eastern Europe and Russia began as a commercial triumph, progressed to an educational tour de force, and reached its climax in the context of the struggle toward détente. On all these levels, just as he had during his New Deal endeavors, Garst embarked on these initiatives as a private citizen.

The visit of members of the Soviet delegation to Roswell Garst's farm in 1955 was among the first interchanges between farmers from Russia and the United States in the years following the Second World War. Henceforth, until his death in November 1977, Garst engaged in a voluminous correspondence pertaining to agricultural matters with officials of Russia and Eastern European nations. During these years the Soviet Union had four premiers or chairmen of the Council of Ministers: Georgi M. Malenkov, Nikolai A. Bulganin, Nikita S. Khrushchev, and Aleksei N. Kosygin. Leonid I. Brezhnev became general secretary of the Central Committee several months before Garst's death and, to provide himself with greater international visibility, assumed the ceremonial post of chairman of the Presidium of the Supreme Soviet of the USSR.

Khrushchev, the Soviet leader with whom Garst is most frequently associated, became premier on 27 March 1958. Contact between the two men was established in 1955 at the time of Garst's first visit to the Soviet Union, when Khrushchev was first secretary of the Central Committee of the Communist Party. At the same time during these years, the United States had six presidents. Garst, therefore, had to consider programs and policies of administrations from Dwight D. Eisenhower on into that of Jimmy Carter. The heaviest flow of his correspondence and of his Russian and Eastern European contacts occurred during the tenure of Nikita S. Khrushchev as the dominant figure in Soviet affairs, roughly from 1954 until October 1964, when he was ousted from power.

The years in which Garst corresponded with Eastern European and Russian officials witnessed marked and dramatic changes in American agriculture, all of which were reflected in his letters. Russian agricultural concerns related primarily to large-scale production in the corn and wheat belt extending from the North Central to the Great Plains states. An agricultural revolution in this region could be said to have occurred in the years following the Second World War, though, of course, its roots could be identified in the prewar years. More corn, for example, was being produced on fewer acres: in 1930 very few American farmers used hybrid seed corn; in 1950 virtually every American farmer raising corn used hybrid seed, which yielded more bushels per acre, from roughly twenty-six bushels per acre in 1930 to thirty-seven bushels per acre in 1950. Because hybrid corn yielded more bushels per acre, additional acreage in the corn-producing states could be devoted to soybeans, thereby increasing the efficiency of feeding livestock: chickens, hogs, cows, and steers. Moreover, as tractors replaced most mules and horses in the two decades prior to the Second World War, more land could be taken out of pasturage and devoted to grain production. By the end of the 1940s farmers were already producing over three billion bushels of corn a year for the highest yields in American history. This figure and many others pertaining to agricultural yields would dramatically increase in the ensuing years as the impact of the agricultural revolution became more and more evident.

Vastly increased use of fertilizers, involving ever-growing quantities of nitrogen in particular, helped increase production, as did extensive use of celluloses such as corncobs and cornstalks in the feeding of cattle, a use that Roswell Garst was prominent in popularizing. Improved technology and better land management also helped expand agricultural production, as did the application of genetic breeding principles to grains and livestock. Enhanced use of contouring, improved drainage systems, and irrigation in areas where these practices were needed further increased agricultural output.

In the postwar decade the new methods became more widespread. Fertilizer manufacturers expanded their production, thereby reducing the necessity for crop rotation, and as the price of corn increased, more farmers in the grain belt began feeding cellulose to their cattle. Garst, emphasizing lower costs, argued convincingly that mixing cellulose with urea, thereby providing necessary protein, would help produce cattle more cheaply. As a result, partly as a consequence of his proselytizing, the demand for urea increased so rapidly that in the early 1950s manufacturers had to ration their output to feed dealers unable to satisfy customer requests. While chemistry increased agricultural production, improved technology did that and more. Less time was required to produce a unit of feed, and planting and harvesting of crops could be accomplished in a more timely and uniform manner. Technology also allowed for more efficient drying of farm products and helped, for example, to provide grains of better quality and higher prices, and in time, after investment costs had been recovered, higher profits. In addition, in the 1950s, agricultural scientists applied the hybrid principles utilized so effectively for corn to poultry, thereby enormously increasing egg production. In 1950, Roswell Garst estimated that 20 percent of all chickens produced in Iowa were bred by these principles. Moreover, the same approach was finding application in hog and cattle breeding.

This agricultural revolution enhancing the ability of Midwestern farmers, whose numbers were constantly declining, to feed an ever-increasing population at reasonable prices was not achieved without cost. Producers, although aware of soil erosion, did not seriously concern themselves with it. Soil fertility, maintained through the continual application of chemical fertilizers, aroused concerns that potential health hazards were being marketed and consumed by unsuspecting individuals purchasing produce in supermarkets throughout the nation. These concerns came to the forefront after the agricultural revolution was well under way. Of more immediate interest was the dilemma of the surplus, which continued to plague American public life in the postwar years in an economy of abundance, just as it had during the Depression years in an economy of scarcity.

However, if grain-belt farmers were to participate in this agricultural revolution, to increase feed and food production as rapidly as population was growing, they would have to expand their operations. Such expansion in turn necessitated costly capital investments under the stimulus of firm prices for produce and government subsidies if prices could not be maintained. But at the outset of the 1950s, with the Korean War sparking demand and with general prosperity thereafter, plus government programs absorbing the surplus or paying farmers not to produce, agricultural prospects were most promising, better than they had been since the golden age of American agriculture, the years before the First World War, when the American farmer functioned on an economically viable base within an expanding American economy and was accepted as a political and social equal on the American scene. By the 1950s, many American farmers in the grain-producing regions and elsewhere were becoming, if they were not already, as efficient and productive—if not so prosperous—as their counterparts in industry.

The agricultural situation in Russia and Eastern Europe was dismally different from that in the American grain belt. Ravaged by war and not yet involved in the revolution affecting American agriculture, production could not adequately meet the needs of its war-weary peoples, whose living standards lagged behind those of the Western world. By the 1960s Russia and some of her Eastern European neighbors were beginning to increase their imports to provide for their expanding populations. Although the soils of the vast grasslands of central Russia and the Great Plains are similar, the climate in Russia is more severe than that of the United States. The growing season is shorter; rainfall is lower on average and subject to greater variability. Because of these factors, added to less advanced means of production and the system of central planning, Soviet yields usually suffered in comparison to American yields.

In all, during these postwar years, the United States produced about 15 percent of the world's wheat; the Soviet Union, with a larger acreage devoted to wheat, produced almost 24 percent. These figures indicate that wheat was not a crop affecting U.S.-Soviet trade and that the Soviet Union at times could export grain, chiefly to countries in Eastern Europe. Such, however, was not the case with corn, the crop that most concerned Roswell Garst. By the end of the 1950s the United States produced over 50 percent of the corn grown in the world, a total of almost 7.5 million bushels. The Soviet Union produced a little more than 5 percent of this total, and Rumania and Yugoslavia together produced roughly the same amount. Russia and Eastern European nations, desiring to increase their consumption of meat, would seek to import seed corn from the United States, despite the fact, for example, that their acreage devoted to grains accounted for more than 70 percent of their total crop acreage in the 1950s.

Though farms in Russia, like those in the American grain belt, were large, they were in the hands of state-controlled collectives or state-owned farms. Although efforts, including those by Roswell Garst, were made, chiefly at Khrushchev's prodding, to improve and modernize farm methods, Soviet collectivist agriculture proved itself more able to supply the state with cheap farm products than to expand productivity, enhance efficiency, and improve the standard of living in the decade following the Second World War. Because of low yields—scarcely half as much per acre as those of the United States—Russia found it necessary to keep more than half of its total crop land in grains, early in the 1960s a total exceeding 277 million acres. During the same period, the United States maintained 160 million acres in grains.

Decision making in the Soviet Union is centralized, and, with regard to agriculture, officials proclaim an overall general policy within the scope of a collectivist planned economy to increase production by large-scale methods to meet the needs of the state. In the postwar decade, mechanization, electrification, and improved farm practices were beginning to make headway on collective and state-owned farms. However, disposal of farm products among domestic requirements, stockpiling, and export demands rested entirely on the decision of officials in Moscow who were not necessarily motivated by commercial considerations. Political factors played a prominent role in Soviet foreign trade operations and could have priority over domestic needs when that was deemed necessary by the government.

In Eastern Europe in the decade following the war, agriculture was in the process of being socialized as the various nations became part of the Soviet bloc. Their farm output was unable to keep pace with increasing domestic demand. In these nations, as in the Soviet Union, industrialization received enhanced emphasis, with some progress consequently being made in the modernization of farming techniques. But, with the sole exception of Hungary, methods of agricultural production in Eastern Europe remained primitive compared to the rest of Europe. Socialization of agriculture in Eastern Europe involved compulsory deliveries to the state and officially determined allocation of material to agricultural producers, with farm and food prices fixed by the state. Foreign trade was state-monopolized and conducted for purposes determined by the government rather than by individual producers.

During the eleven years Nikita Khrushchev was the dominant political figure in Soviet life, his agricultural policies, emphasizing the extensive purchase of seed corn and parent stock to the detriment of wheat and other crops, helped to bring Roswell Garst to the attention of Soviet officials concerned with agriculture. Khrushchev, who repeatedly complained that Soviet agriculture did not produce enough food and livestock feed, emphasized corn as the most efficient fodder for an increase in meat and milk production. After his fall from power, Khrushchev's successors quickly began to reverse this emphasis on corn and called for an increase in the production of equipment and other agricultural inputs. As Khrushchev's corn campaign, during which the area in the Soviet Union devoted to corn increased almost tenfold, came to an end, agricultural officials at all levels had a chance to present proposals related to local conditions. After 1963 Garst's relations as well as his impact on Soviet agriculture waned but did not come to an end. Although corn production declined and a different balance in the political economy of Soviet agriculture was sought, Leonid Brezhnev had to remind some officials that corn was still a valuable crop, although it was no longer "the queen of the fields." Forced in 1963 to import wheat because of poor yields, the Soviet Union after Khrushchev sought to expand production of winter wheat, and many of the new lands formerly devoted to corn were gradually allowed to return to pasture for grazing purposes.

Though Soviet production of corn increased dramatically under Khrushchev, it never seriously challenged American output. In 1961 the area devoted to corn in both countries was about equal, but in volume of output Soviet production was about a quarter of that in the United States. And 1961 was the best year for Soviet corn yields. Khrushchev's concern for corn production becomes understandable when one realizes that it was a relatively new crop in Russia before the revolution and that during the quarter-century Josef Stalin dominated Soviet life its production remained static. The acreage devoted to it throughout these years was less than what it was in 1928 when Stalin firmly assumed power. Moreover, advances in corn production, such as the use of fertilizers and hybrid seeds, were not widely utilized until Roswell Garst proclaimed and demonstrated their virtues to Soviet farm leaders. Once Khrushchev secured access to hybrid corn, it was planted everywhere: the virgin lands of Kazakhstan, the cold north, areas near the Black Sea, as well as the outskirts of Moscow. In all, from 1950 to 1960 the Soviet Union increased its crop land by 141 million acres, 100 million acres of which were lands in Asia. In 1961, it had a total of 505 million acres in crops, compared with 310 million acres in the United States.

Much of the corn raised in Russia, unlike that in the United States, was used for fodder and not harvested for its grain, in part because most of the Soviet Union lies within the same latitudes as Canada; the Crimea, for example, is in the same latitude as Minneapolis. When Khrushchev was removed from office in the fall of 1964 by the Central Committee of the Communist Party, agricultural production had increased only 10 percent, a far cry from the 70 percent that his program envisioned.[2]

His successors, Kosygin and Brezhnev, as previously noted, more effectively than Khrushchev, emphasized increased production of farm implements and widespread use of fertilizers. Irrigation projects were expanded; huge sums were devoted to these ventures in a massive effort to expand production through the application of modern science and technology, as well as through revision of management techniques on state and collective farms by allowing greater initiative on the farms, with the goal of enhancing their self-sufficiency and expanding their production.

These measures helped expand production. In 1966 and 1967 the

---

2. Information and data about Soviet agriculture were derived from Naum Jasny, *Khrushchev's Crop Policy* (Glasgow, George Outram & Co., n.d. [1964]); a pamphlet by Lement Harris, *Farm Peasantry to Power Farming: The Story of Soviet Agriculture* (New York, International Publishers, 1968); and Martin McCauley, *Khrushchev and the Development of Soviet Agriculture* (London, Macmillan, 1976). In addition, many factual data were derived from a June 1964 confidential report prepared for the National Planning Association, "Farm Trade with the Soviet Bloc," by Ralph Yoke; a copy is available in Box 5 of the Donald Murphy Papers, Archives of American Agriculture, William Robert Parks and Ellen Sorge Parks Library, Iowa State University.

Soviet Union enjoyed the largest grain crops in its history, based largely on yields of wheat, rye, and barley—but not corn—along with bumper yields for other major crops. Nevertheless, as under Khrushchev, state purchases of grain continued in years of poor crops, and corn was still a matter of concern. After Khrushchev Soviet agricultural and other officials, though on a reduced scale, maintained their contacts and correspondence with Roswell Garst until his death in 1977.

Whereas the political economy of Eastern Europe directly determined the nature of its agriculture by defining its direction, setting its goals, and formulating its policies and programs, in the United States the problem that plagued agricultural officials and the government itself was the dilemma of the surplus, the result of initiatives by farmers, manufacturers, scientists, agricultural economists, government bureaus, agricultural colleges, and others, all of whom played a role in improving and expanding agricultural output in the postwar years. (This dilemma, of course, was not one that plagued agriculture in Eastern Europe.) Roswell Garst was prominent in helping to expand American agricultural production, but being neither a public official nor a politician, he did not accept it as a dilemma. He saw enhanced production as an opportunity for fewer farmers to feed rapidly increasing numbers of people both at home and abroad. And given the opportunity, he was determined to assist and guide officials in other countries in ways and means, based on his experience, of improving their production of feed grains. In doing so, Garst needed the approval of American officials trying to cope with the increasing agricultural glut in a country that was involved in a cold war with the Soviet Union and other communist bloc nations.

In the postwar era the farm problem was one of the major issues of American public policy. At its base was the phenomenon that production, despite a distinct and steady decline of farm population, increased more rapidly than demand and thereby created constantly accumulating surpluses. In congressional debate and in presidential campaigns, the issue produced sharp sectional and party clashes. In 1977, at the time of Garst's death, it was as unresolved as it had been in 1945.

The surplus, the central theme of the farm problem, resulted from increased specialization and application of science and technology to agricultural production. The agricultural revolution of the postwar decade involved the acceleration of tendencies that had slackened during the Depression years but had started up again during the war years and expanded markedly in the succeeding decades, a theme previously discussed in this essay. Its chief result, aside from sharp increases in productivity, was a noticeable trend toward the most up-to-date technical advances. During the years of the Second World War on through the Korean War, from 1941 through 1953, the vast increase in agricultural productivity was balanced by a demand that provided ready and favorable markets for farm goods.

With the end of the Korean War, just as Garst's initial meetings with Russian officials were getting under way, demand declined or remained static while production continued to increase. Large surpluses began to accumulate. What to do about them was the question that, since the 1920s, had both intrigued and plagued farmers and their representatives in state houses and in Congress, as well as agribusiness officials, agricultural economists, and others involved with the agricultural and rural sectors of the American economy. The paradox of plenty, during the years in which the cold war waxed and waned, would provide part of the backdrop for Garst's Russian and Eastern European contacts. The other part, already reviewed, was the agricultural situation in Russia and Eastern Europe.

Efforts to resolve the dilemma of the surplus, the paradox of plenty, projected agriculture as an important public, hence political, issue in the 1950s. In the 1940s, during the war years and shortly thereafter, the government sought to encourage production by guaranteeing higher prices to farmers for their produce. But by the end of this decade, and before the advent of the Korean War, when postwar agricultural demand began to slacken, powerful arguments favoring a transition to a free market were presented.

Proponents of this view insisted that huge surpluses would allow prices to fall, thereby resulting in reduced production because less efficient farmers, unable to operate at a profit, would leave farming. Meanwhile, the advocates argued that remaining farmers, because of lower prices, would have to reduce capital investments and production inputs, further decreasing production until supply and demand reached a rough equilibrium. A free market, in short, would automatically end surpluses and drastically reduce, if not end, the need for government expenditures in pegging market prices.

Opponents of the free market claimed that such a policy, besides collapsing farm prices, would be responsible for the economic devastation of thousands, possibly millions, of farm families. The consequences could involve a national depression thrusting vast numbers of people to urban labor markets and unemployment rolls, as well as the destruction of traditional social values supposedly embedded in American family farm life. Federal support of farm prices, along with programs and policies to sustain smaller operators, plus production controls to curb the accumulation of surpluses, this group argued, was the best way to handle the paradox of plenty. In the years during which Garst corresponded with Eastern Europeans, Presidents Truman, Kennedy, and Johnson endorsed this posture.

Ranged between these two positions, advocating a free market or a carefully managed agricultural economy, were individuals and groups whose opinions combined features of both points of view. President Eisenhower found many members of Congress in accord with his position of endorsing a low level of income support to protect farmers from drastic drops in market prices but opposing income guarantees regardless of long-term demand and price trends. However, in the three decades following the Second World War, for the remainder of Garst's lifetime, a consensus for a comprehensive permanent policy of either a free market or a carefully managed farm economy could never be reached. What prevailed were shifting compromises involving party, regional, and commodity interests, as well as cold war factors that resulted in farm bills that sought to protect income at reduced levels and to curtail production, although not very stringently.

Two types of programs for sustaining farm income were devised in the postwar years. One did not directly affect the market; included in it were activities such as cheap credit, technical advice, assistance in securing water supplies, soil conservation, and disease and pest control. The other called for activities and controls to shore up farm income by sustaining high prices; this was to be accomplished through price supports, disposal programs, and production controls.

In supporting prices the government stood ready to take off the market at a given price, within a range determined by legislative action, supplies of specified crops. This approach was designed to maintain the market price at the level of the support price. The specific support price was usually determined by the secretary of agriculture within limits set by Congress. It involved crops responsible for the greater production of farm income, including those that concerned Garst such as corn, wheat, small feed grains, and soybeans. From the end of the Second World War to the end of the Korean War, supports for these and other major crops were maintained at 90 percent of parity. Support floors during the remainder of the Eisenhower administration ranged from 60 to 75 percent of parity, depending upon the commodity involved. Though they continued to fluctuate with succeeding administrations, they never again reached 90 percent of parity. President Johnson in 1965 won congressional approval for a revision of production limitations and price support devices. As a result, farmers were to set aside a portion of their crop lands from production; in the case of grain farmers the specified crops were corn and wheat. If these lands were devoted to soil-conserving uses, the farmers were eligible for price support plus special direct cash payments. Though the immediate effects were reduction of surpluses and heavy donations to hungry people overseas, the program aroused partisan differences never fully resolved during Garst's lifetime.

However, the notion of assisting hungry people overseas as a means of getting rid of agricultural surpluses was an attractive one, and to no one more so than to Garst. Inaugurated in 1954, the Food for Peace program (PL 480) initially emphasized building markets for surplus agricultural commodities; its aims expanded in the mid-1960s to programs of combatting malnutrition and hunger and promoting economic growth in developing nations, along with its established role of expanding export markets for surplus American agricultural production. Through June 1972, agreements covering sales on credit of commodities with a market value of about $13 billion were reached with fifty-five countries. In addition to these sales through the Food for Peace program, by the end of June 1972, produce valued at $3.5 billion was sold on long-term credit, and additional agricultural commodities were donated to assist in meliorating famine and other urgent relief needs overseas. Moreover, Congress, once cold war tensions abated somewhat, was willing to sell surplus grain to the Soviet Union.

The embargo on Soviet grain sales had been broken almost a decade earlier, during the Kennedy years, following the 1963 test-ban treaty. In October 1963 Kennedy, after several weeks of frantic behind-the-scenes maneuvering, announced that export licenses would be issued to allow American grain traders to deal with the Soviet Union. To be sure, prior to 1963, limited trade existed with the Eastern European nations, but cold war tensions and anti-Communist outbursts kept such trade at low levels. And it was not until 1972 that officials of the American government negotiated an agreement with Soviet officials. Just as corn was politics in the Soviet Union, so trading with the Soviet Union was politics in the United States. While U.S. surpluses mounted, Canada, Australia, and other grain-exporting nations enjoyed a market in Eastern Europe and Russia free from American competition. With the easing of cold war tensions, grain sales could be conducted with reduced friction and hostility, such as American longshoremen's refusing to load grain destined for Soviet ports.

In July 1972 the Soviet Union and the United States agreed to a historic grain credit arrangement. The U.S. government granted the Soviet Union a $500 million line of credit in return for a Soviet pledge to purchase, over a three-year period, $750 million worth of American grain. The renewal of the agreement ran into difficulties when the Ford administration accepted stipulations supported in Congress tying grain sales to a modification of Soviet emigration policy. The Soviet government repudiated these terms, and they were modified, so that in 1976 a five-year Soviet-American grain agreement was signed in Moscow, with the Russians pledging themselves to purchase six to eight million tons of grain each year. Soviet purchases involved wheat more than corn, though at the outset Americans thought that corn would be the commodity Soviet officials most desired. Garst's efforts in promoting Soviet corn production undoubtedly had some influence in shifting the emphasis of Soviet grain purchases.

No sooner had trade restrictions eased than Khrushchev, in October 1964, was ousted from power, leaving Garst without his chief contact in the Soviet Union. Though his contacts with Soviet officials continued for the rest of his life, at the time Garst thought they most certainly would come to an end. However, he soon found that the Soviets wished to continue their contacts and that American officials were equally eager that Garst maintain them and continue his visits to Russia and Eastern European countries. In all, beginning in 1955, Garst made seven trips to Russia and Eastern Europe and entertained scores of Soviet and Eastern European agriculturalists at his Coon Rapids farm.

Like those of F. Hector St. John Crevecoeur in the Eighteenth Century, Garst's letters of the midtwentieth century present to a wide audience a generally sympathetic, yet clear-thinking portrait of "the nature and principles of our association" in the United States. Both Crevecoeur and Garst in their letters explained the nature of American agriculture in terms of their respective experiences. Both enjoyed farming, were fond of their neighbors, and resided in rural villages, Crevecoeur for several years in Pine Hill, New York; Garst in Coon Rapids, Iowa. In the Introductory Letter to his *Letters From an American Farmer*, first published in 1782, Crevecoeur wrote to the Abbé Raynal that "the treatment you received at my house proceeded from the warmth of my heart and from the corresponding sensibilities of my wife: what you now desire must flow from a very limited power of mind: the task requires recollection, and a variety of talents which I do not possess. It is true I can describe our American modes of farming, our manners, and peculiar customs, with some degree of propriety because I have very attentively studied them; but my knowledge extends no further."

Two centuries later, Roswell Garst could very well have written similar words to one or another correspondent in Eastern Europe or Russia, but with one modification: as a supersalesman for American agriculture, he would not have deprecated his abundant talents.

The selection of letters appearing in this volume are part of the vast collection of Roswell Garst papers that constitute a portion of the Archives of American Agriculture housed in the Department of Special Collections of the William Robert Parks and Ellen Sorge Parks Library at Iowa State University. When fully processed, the Garst papers will run to more than one-hundred boxes. Their subject matter includes not only Garst's business interests and the material represented in this volume, but virtually the entire range of topics and issues in midwestern American agriculture from the 1930s through the 1970s, including farm management, technological developments, agricultural economics, farm programs, and a variety of correspondence with leading national and international figures of the time. The letters are filed chronologically, alphabetically, and topically. Though box numbers are not cited, researchers interested in pursuing special topics or particular individuals will have no difficulty in locating materials with the assistance of the Roswell Garst inventory or shelf list.

We wish to thank Laura Kline and Becky Jordan for their assistance in helping us track down various letters while the collection was being processed. Dean Warren Kuhn and his staff in the William Robert Parks and Ellen Sorge Parks Library also merit a vote of thanks for making the library a most congenial and conducive place for scholarly endeavors. For expert assistance in preparing this manuscript we are indebted to Audrey Burton, Mary Cretsinger, Curtis Chrystal, and Tom Slaybaugh. And for their continuous support of scholarly research in the humanities at Iowa State University, Richard Lowitt is indebted to Daniel Zaffarano and Norman Jacobsen, respectively dean and associate dean of the Graduate School. Finally, we are most grateful to the members of the Vigilantes who carefully critiqued the introductory essay. Though we did not accept or agree with all of their comments, all were carefully considered.

Correspondence with officials in all Eastern European countries is included in this volume, although, as would be expected, the bulk is with Soviet officials. Outside Russia, Garst's most extensive contacts were with Hungarian and Rumanian officials. His letters, since he was seeking to influence people, were usually lengthy and, over time, repetitive and excessive in detail. Consequently, we have deleted repetitive details and extraneous materials from the letters in this volume. Deletions are indicated by the standard ellipsis notation [. . .]. Since Garst dictated directly to a secretary and rarely proofread his letters, we have, when necessary, corrected typographical errors, grammatical errors, spelling, and punctuation. In the case of letters to Garst, we have made fewer changes, but have corrected obvious typographical errors and some punctuation, when required for clarification of meaning. Additions in both categories are indicated by brackets. In both types of correspondence, we have attempted to preserve the flavor and appearance of the letters as much as possible, but at the same time have attempted to make them more easily readable and coherent.

Letters to Garst were usually brief and perfunctory, since many of his letters, analytical and exhortatory in nature, did not call for a response. Hence few letters to Garst appear in this volume. Of the 104 letters published, only 26 are to Garst. Letters Garst wrote to American officials and friends are included because they either summarize or further illuminate his Russian and Eastern European activities.

Each set of letters is preceded by an introduction placing the correspondence in more meaningful perspective. Footnotes identify the few officials Garst did not identify himself and at times further elucidate references within the text of a letter. We have adopted a "to-from" system for the Contents pages and for the letter headings in order to make both the sender and the recipient immediately identifiable to the reader.

The Appendix, an article written by Garst in his letter-writing style that appeared in *Pravda* and *Izvestiia,* was requested by Khrushchev and cited by the Soviet leader at the Central Committee meeting in February 1964. The article made Garst's views, as well as his name, known to a wide audience throughout the Soviet Union.

Letters from an
American Farmer

EASTERN EUROPE

Garst's cordial reception in the Eastern European countries in 1955 and thereafter sprang from his successful tour of the Soviet Union and his first meeting with Nikita Khrushchev. However, Garst had determined from the very first to visit the corn-producing countries of the Eastern bloc as well. His success with two of them, Rumania and Hungary, was among his major contributions to the East-West dialogue. His correspondence with officials in all of those countries reveals his persistence in promoting his product, the energy he expended in giving freely of his knowledge, and the way in which he established personal friendships during an era of deep mutual suspicions and misconceptions between East and West. The manner in which different individuals and different governments responded to his approach is also consistent with the variety of national responses, within an ostensibly monolithic political and economic structure, to the new opportunities presented to Eastern Europe.

Rumania, Garst's first stop after Russia in 1955, plunged him most quickly into intimate professional and personal relationships and brought forth his first direct appeal to an Eastern European country for improved diplomatic relations with the United States. All of his themes were welcomed in Rumania, traditionally the largest corn grower in Eastern Europe, eager to develop the new genetics of hybridization, determined to modernize all of their agricultural practices, and anxious to repair a hitherto antagonistic relationship with the United States and the West. Rumania's success in these fields and Garst's contribution to the transformation of their feudal agriculture are vividly revealed in the letters selected for this section. Only two years after Garst's initial visit to the country, his friend Gabriel Reiner of Cosmos Travel, New York, visited Rumania and spoke to Prime Minister Stoica Chiou. "We talked quite a bit about you," Reiner reported to

Garst, "and it seems, without question, your name is fixed in their history, regardless of politics, in honor of your great contribution."

Garst did not visit Bulgaria, Czechoslovakia, Yugoslavia, and Poland in 1955, but initiated correspondence with these countries and entertained delegations from each of them during the late fifties and early sixties. He made brief stops in Czechoslovakia in 1955 and 1957, Yugoslavia in 1957, and Bulgaria in 1959. His commercial success and personal impact were less pronounced in these countries, but his persistent attempts to sell seed corn and to be helpful as educator and commercial agent did produce some results.

Bulgaria is perhaps the best example of Garst's sheer doggedness: a refusal until 1959 by the State Department to allow him to travel to Bulgaria or entertain delegations caused problems, but eventually, after a visit to Bulgaria by his son David and torturous negotiations in New York, Garst achieved a sale and introduced Bulgarian officials to American officials. Yugoslavia, distancing itself from the Soviet bloc, bought a small amount of seed and asked for other help. Czechoslovakia, although it grew corn along the Danube near Hungary, was recognized by Garst as more highly industrialized and consequently a better market for machinery. Poland, the only country courted by Garst that was not associated with the corn-growing area along the Danube and its basin, sent its American ambassador, and later a delegation, to Coon Rapids but did not follow up this contact with Garst. It should be stressed, however, that Garst's correspondence with all these countries, even Poland, is extensive, and that much of it has been deleted because it repeats themes already expressed or developed more fruitfully further on in this edition.

During his first trip in 1955, Garst visited Hungary, a country whose officials, like those in Rumania, responded immediately and enthusiastically to Garst's personality and the prospect of acquiring Western technology. Sandor Rajki, newly appointed head of the Agricultural Research Institute of the Hungarian Academy of Sciences, decided to introduce, under Garst's tutelage, hybridization of Hungary's entire seed corn production. To achieve this end, Rajki built a large processing plant at Martonvasar and sat down with Garst to mark the location of the thirteen regional plants later constructed. Early sales and information gathering from Garst were interrupted by the Hungarian uprising in October 1956, but contacts and agricultural progress gathered momentum again in the sixties. In 1977, Hungary began its first large-scale experiments with the Garst program of feeding protein-enriched cellulose to cattle.

We begin the correspondence with Garst's account of his first visit to

Russia, Rumania, and Hungary in the autumn of 1955. The letter constitutes an appropriate prologue to all that follows and conveys the excitement Garst felt at the extraordinary opportunities that lay ahead.

November 2, 1955

To Fred Lehmann, Jim Wallace,
Bob Wood, Hugh Morrison[1]

Gentlemen:

I am going down to Des Moines tomorrow to tell the Pioneer of Des Moines group about the trip, but I thought I'd just as well write it down for you all because it was quite an astonishing total story.

We got to Moscow September 25. For the information of Bob Wood and Hugh, the "we" means Dr. Geza Schutz, a longtime friend of mine who was born and raised in Hungary; got his Ph.D. from the University of Geneva; took further studies in France, England, and Columbia University, all in economics; speaks German, Hungarian, French, Italian, and Spanish and all of them relatively fluently; a man who had written his thesis for his Ph.D. on the subject "The Condition of the Worker and Peasant in Central Europe from 1890 to 1914" and who is at the same time an accomplished musician. He could play the Russian folk songs, the Rumanian folk songs, the Hungarian folk songs on an accordian, a piano, or a violin, and they all love music. He knew their history—he was without doubt exactly cut out as the best possible companion for me. He furnished the culture and I furnished the agriculture. He actually is presently a farmer up at Janesville, Minnesota, although not too seriously. He is on the faculty at Shattuck Military Academy—teaches philosophy—and spent ten years as a labor conciliator with the U.S. Department of Labor so he also knows how to negotiate. He just is a top guy and is a top guy in many respects, and we have been friends for fifteen years so he knew the whole background.

We spent five days in Moscow. We first submitted a very long letter to Mr. [Vladimir] Matskevich, for translation and study, and we spent two days out at their Agricultural Exposition, which is a very grand exposition. There are sixteen Socialist Republics in the Soviet Union—about as we have states only, of course, very much larger than our states. Each one of the republics has a building at the Exposition, and then, of course, they have several central buildings that show the exhibits of all of Russia and

1. Officials of Pioneer Hi-Bred Corn Company.

such things as machinery and industry. We also spent one day out at an experimental station near Moscow and visited with all of their scientists—looked over the university and visited the collectivist farm. Then the last couple of days we were there we spent mostly around the Ministry of Agriculture. They would bring in twenty or thirty of their top men to question us on the mechanization of American agriculture, on exactly how we produced hybrid seed corn, on the necessity of drying, the temperatures of drying, the equipment required. They asked us about insecticides, fungicides, methods of conducting yield tests, etc., etc., etc. Mr. Matskevich put on a very beautiful luncheon for us with the top end of the agriculture department, but then mornings and afternoons for two days we would spend two or three hours answering questions.

Mr. Matskevich, who is now minister of agriculture, told us that they definitely would want to buy American hybrids from us but that they had not arrived at the amount and they were anxious as I was to have a study of maturities. So Friday evening we flew to Kiev and we spent roughly two days each at Kiev, Kharkov, Dnepropetrovsk, Odessa, and Kishinev. We would see a collectivist farm and an experiment station and a state owned farm and a tractor station in each location and meet with all of the leading people at each location, and it always called for some kind of meal even though we had had a meal two hours earlier. And always with lots of wine, and always with lots of toasts—and just 100 percent of the time the toasts included the words "peace" or "the spirit of Geneva" or "increasing friendliness between our two countries" or any other set of words that added up to the same thing. They would always give a little speech but end up with a rather short toast at the end of it. I soon found out how to do it—I would point out that if it was sensible to trade information on the peaceful uses of the atom as was done by all countries at Geneva recently, then it was equally sensible for the world to trade information about how to produce more food—especially meat, eggs, and milk—and that meat, eggs, and milk were best produced through the use of corn; that the trading of information about the peaceful uses of atomic energy have brought the countries of the world closer to the peace that we all so much desired; and that Dr. Schutz and I were primarily corn people and that I would therefore propose a toast "peace through corn." That was my standard toast for three weeks if I only needed to make one. I would vary it a little bit if I was forced into two.

In a by and large way the first day or two were confusing because they kept talking about this area being planted to 25 percent hybrid corn and the next area 30 percent hybrid corn, but it all looked like open pollinated corn to me. Finally, I found out that when they crossed a dent variety on a flint variety they called the resultant product hybrid corn and that we were talking about two different things. And I will say that they got a big improvement in their corn by making such a cross. Originally, most of the corn was flint. They discovered that crossing the dents on the flints improved their production so they have been doing this on quite a large scale.

In a broad way, I would say that this cross bred corn which is now used on approximately one third of their acreage is as good as the open pollinated corn that existed in the corn belt when I first went into the hybrid seed corn business. Probably more than one third of their total corn acreage is now planted with this kind of seed and two thirds is still planted with flint.

This was not only true where they need early and extra early varieties but was also true down in the Odessa area where they had big eared, shallow kerneled flints with a stalk of good big stature and where their cross bred corn grew with as great a plant stature as Pioneer 302 in good bottom Missouri land. I think, of course, that good American hybrid seed will increase their yield over this cross bred corn by 30 percent, and I think it will frequently be 50 or 75 percent better in yield than the original flints, but they are making some progress. Of course, at every experiment station and research station they knew about four-way crossed American hybrids, exactly how they were produced, and all of them realized that the four-way cross will bring them an increase in yield and a very great improvement in stiffness of stalk and much greater uniformity and will permit them to mechanize.

One of the startling discoveries I made was that practically every acre of corn is planted in seventy-centimeter rows, which is roughly twenty-eight inches. They have a great many tractors in Russia, all of them practically the same model—a big strong Caterpillar diesel that will pull five or six sixteen-inch plows. They also have big combines, and it looked like pretty nearly enough of them. They hook as many as five grain drills behind these Caterpillars so they have their wheat and rye production reasonably well mechanized. Furthermore, they plow all the ground for corn with these big tractors and disc it with them, and then they pull two six-row planters and two six-row cultivators. Of course, they can only cultivate corn to about knee-high because the clearance is low on both the Caterpillars and the cultivators. They never did hear of 2-4-D or DDT, but they didn't seem to have too many insects and they have plenty of human beings to hoe out the weeds after they get through cultivating. The horrible part comes at picking time.

They have too few horses and wagons and too many people so they can't afford to have the corn picked directly into the wagon. They pick it and throw it on the ground and then they either pick it up off the ground with baskets or directly into the wagon. Then they take a corn knife and a woman and she cuts the corn off at the ground, that is, the corn stalk, and piles it in a little pile. Then they tie it in little bundles with slough grass and subsequently shock it or haul it directly to the house or farmyard. They spend a terrific amount of labor at harvest, but they save every stalk finally and they save all the straw from the grain. They raise no soybeans at all, but they do raise thousands of acres of sunflower, which they use as their oil crop, and they use sunflower meal for some of their protein. However, they have far too little protein so they raise a great many mangel beets and pumpkins and squash and even watermelon for cattle feed. If they are feeding dry corn stalks or dry straw, they supplement it with beets, small potatoes, pumpkins, squash, watermelons, etc. They even mash these up and make silage out of pretty dry corn stalks by putting a layer of corn stalks and a layer of pumpkin, then a layer of corn stalks and a layer of mangels, then a layer of corn stalks and a layer of watermelon, and they say that the moisture provided by the other things transfers itself into the corn stalks and that it makes pretty good feed. They raise very little hay, but they do feed all their straw and stalks.

Some few years ago they discovered that they could pick the corn at the waxy stage as they call it and grind up the ears, shucks and cobs and all, and put them in one silo and then chop up the stalks and put them in a different silo and have as many total feed units—or so they believed—as they would get out of mature corn. They feed the ground ear corn silage not only to their hogs but to their geese, ducks, and chickens. I did not see any of it being fed, of course, but we did see it being made and it looked like excellent hog feed to me, and I thought ducks and geese might do well on them but I didn't think too much of it for chickens.

When we were at Odessa, which is on the Black Sea, they informed me that Mr. Khrushchev would like me to come over to Yalta, to which of course I agreed. Because Mr. Matskevich had not invited Geza Schutz to come to Russia, he was not invited to come to Yalta to see Mr. Khrushchev although he had been very well accepted everywhere. We couldn't figure it out except that they thought they could work on one man better than two in the purchase of the corn, which was done at Yalta.

It is two hours by plane from Odessa to the Crimea—and two hours by car through a very rough, mountainous country from the Crimean landing field to Yalta. We arrived at Yalta about 10:00 A.M., were taken to a guest house to rest and change our clothes, and the interpreter and I were then taken over to Mr. Khrushchev's home about 2 o'clock in the afternoon. Mr. Khrushchev and Mr. [Anastas] Mikoyan, their secretary of foreign commerce, and an interpreter, and I visited from 2 to 4 and then came Mr. Matskevich, [accompanied by] the secretary of agriculture of the Ukraine and several of Mr. Matskevich's deputies from the Department of Agriculture, and the conversation went on for 2½ hours more. Mr. Khrushchev is the head of the Communist party, which is now supposedly run by a top committee of which he is one of the most influential members. He is the man that said last February that Russia ought to have a corn belt like Iowa and produce more meat, which led to the interchange of farmers between the United States and the USSR.

So, he has been the most insistent one on corn. Everywhere we had been, we had been told that Mr. Khrushchev wants more acres of corn and that, therefore, the corn is being expanded. That was everywhere. They connect the two things, Khrushchev and corn, completely.

Habitually, Russia used to raise about six million acres of corn before they took Bessarabia as it was known when it belonged to Rumania. Bessarabia always raised about 2 million acres of corn so you can say that the present Russian area used to raise eight million acres of corn. They claim that by 1954 they were up to twenty million acres, by 1955 they were up to forty million acres, and all of their goals called for sixty million acres by or before 1960. I, frankly, doubt that they have gone anything like that fast although there is no doubt at all that the corn acreage is expanding. I suspect that when Mr. Khrushchev says "plant forty million acres of corn" Mr. Matskevich tries to get that much planted, but it's difficult actually to get it accomplished and that they have not quite accomplished as much as they claim. But they will one of these days.

Mr. Khrushchev and Mr. Mikoyan both told me that the primary need in the USSR is more high protein—meat, milk, and eggs; that they can produce plenty of wheat, potatoes, and cabbage, and rye; that they are not hungry—and this was true. But the more industrialized they become, the more important high protein food becomes. A peasant on a farm can eat lots of borscht—cabbage potato soup with a little bit of meat and some carrots that is, incidentally, very delicious. But it's hard for an industrial worker to eat bread without cheese or meat.

They have decided as a result of quite a few experiments that there are three or four times as many feed units in an acre of corn even picked at the waxy stage and ensiled as there is in an acre of oats matured. They don't really care too much about the maturity—they want a decent-statured plant that can be harvested at the waxy stage and ensiled in two different silos— the stalks in one for the cattle and the ground ear corn in the other for the swine, chickens, geese, and ducks.

Khrushchev told me that Matskevich and his deputies would order the seed corn the next morning but that he wanted to find out about getting some inbreds or some single crosses for the production of hybrid corn within Russia itself and that he wanted to send over engineers to study our drying, grading, treating, and bagging preparation methods and that they wanted to get into the actual production of hybrid seed as soon as possible. I told him that college inbreds would be available and that our geneticists would be able to help their geneticists select the best end of the college inbreds; and that I always thought it was possible that the Pioneer organiza- tion might be willing to furnish them with some single crosses, but that I thought privately owned inbreds would hardly be available, although this was not my department and I could only discuss it with them upon my return, etc., etc., etc.

Incidentally, after this long conference of something like 4½ hours he said that that was enough for one day and we would knock off and look over his beautiful place and then have dinner. Mrs. Khrushchev and their daughter gave me a personally conducted tour of their beautiful formal garden, swimming pool, informal gardens, and their view. Incidentally, the interpreter furnished me was a woman all the time I was in Moscow and all over the countryside including the visit to Yalta. We started eating about seven and I would say that the dinner lasted until well after nine. It was of course very elaborate and very elegant and very delicious and very beauti- fully served as you would expect with the finest wines from the Ukraine, from Georgia, from Uzbek, which is one of the republics lying southeast of the Caspian Sea, the pictures of which look like southern California.

The next morning I returned to the office part of the Khrushchev residence and visited with Matskevich and Mikoyan and the deputies from the agriculture department, and they placed a firm order for fifteen hun- dred tons of extra early corn or two thousand tons, fifteen hundred tons or two thousand tons of early maturing corn, and a minimum of one thousand tons of medium maturing corn, stating that the minimum order would be forty five hundred to five thousand tons and that they probably would end up by increasing above that minimum. They do not want any Large Round—which I didn't think was too serious—and they whined a good deal about my not letting them have any Medium Flat. However, they didn't much mind the Medium Round and they didn't mind the Small Flat.

As a matter of fact, our Small Flat and Short Medium Flat are better kernel sizes than they have always planted.

They said they would send a delegation over directly after my return to conclude the purchase because I didn't at that time know what varieties or kernel sizes we have in long supply and I could not be very specific. Their delegation is going to consist of one geneticist—one of their best engineers to study the machinery necessary for the production of hybrid seed corn and also to study the most modern farm machinery for the commercial production of corn and potatoes. Then they are sending a livestock expert, to study the use of urea and molasses as protein for their cattle feeding, and an interpreter. That makes four in all. I told them I would not arrive home until about now, but I have not as yet received word as to the exact date of their arrival. I don't think it will be too long now.

We flew back to Odessa, spent two very interesting days in the Kishinev area, which used to be called Bessarabia and was a part of Rumania and is now called Moldavia and is part of Russia. It's the best corn country in Russia or at least one of the best.

Before I leave Russia, I must tell you that it is completely communized and has been for such a long time that communism is completely accepted. They own every acre of land, that is, the state does. They own every small retail business. Even the woman selling lemonade in the park works for the USSR; every store is state-owned. The only private employment is babysitters and maids. It is just hard to get used to it and I am sure it is terribly inefficient. But they are used to it and they think it's grand. Catholics in the United States eat fish on Friday and they don't object—they have always done so—and I even do it myself quite frequently. It has always been part of their life. That's the way communism exists in Russia. It's been there forty years or thereabouts—it's better than they used to have with their feudal systems—and they are not objecting. They are as proud of being party men as a Baptist is of being a Baptist. It is hard to explain and amazingly different and terribly inefficient but it is not going to be revolted against.

Anyhow, lets get down to Rumania, which was even more interesting than Russia. We took the train from Kishinev to Bucharest and arrived in Bucharest one evening about 8 o'clock. We had obtained our Rumanian visas while we were in Moscow, and the Rumanian counsel in Moscow had specifically asked us to let her know on what train and in what car and in which compartment we would arrive in Bucharest so the government could receive us, which we had done. We were met by the president of the Chamber of Commerce, which is of course part of their government; by the first deputy minister of Agriculture; by the leading editor of the newspaper, who is also a member of their Congress; and by their leading geneticist and the leading engineer of the Ministry of Agriculture.

I presumed that some of this was due to the fact that the whole press of the Soviet countries had carried full information about my being entertained by Mr. Khrushchev a few days earlier, but anyhow it was a real welcome.

They took us up to the guest house of the government, which would be the equivalent to taking us to Blair House in Washington, D.C., which is the place that all of the visiting royalty is housed by our own government. We were given time to wash and clean up and then sat down to a very fine dinner and a very protracted visit which lasted until about midnight and during which they outlined what they were anxious to show us, the total trip being for twelve days. We told them we didn't dare spend more than eight days because we had invited the Russians over and were going to have some of them over and were going to invite some Hungarians over and so we had to be home to act as hosts.

The next day we saw Bucharest and their fine experimental and research center for agriculture, which is only a few miles out of town; saw their beautiful parks; had a private showing at their art museum after dinner that night, etc., etc., etc. The whole group that originally welcomed us were with us for the next eight days.

The second day we started in the morning in a private one-car train, a very long diesel-powered outfit with berths for fourteen people, full dining and lounge car facilities, and with literally 150 feet of red carpet stretched from the curb where we left our automobiles through the depot to the special car. The depot was one on the edge of town which had been built by the king for his private use. We spent the next three days looking [at] their corn fields, their experiment stations, their collective farms, their state-owned farms and the terrific problems they have not solved—the small peasant farms. The state only owns 26 percent of the land in Rumania, most of the rest of it belonging to peasants who have five to fifteen acres each and even worse, the five to fifteen acres that they do own may be in four different patches three directions from the village. They don't know how to handle that one and neither do I.

Even on collectivist farms it is tough enough. You see, the Rumanians were under the heels of the Turks for five hundred years. They fought the whole five hundred years, first with Russia on their side and then with Austria.

They finally got rid of the Turks in 1848 but continued on under a system that was at least equally bad of kings, archdukes, dukes and "boyars" who were large landowners. It was completely feudalistic with no education for anyone except the very few at the top and with the lowest possible standard of living for all the peasants and laborers. Dr. Louis Michael, who taught Henry Wallace chemistry at Ames fifty years ago and who spent some twenty-five years in that general part of the country, told me after I returned that it was about as bad as it was possible to describe even when he was there, which was mostly from 1910 to 1935. They went through the First World War and its reconstruction period and probably had a few years when they were gaining during the twenties and early thirties. Then Hitler encouraged the Rumanian Fascists to seize control—and even before the war sent German Fascists. And during the war it was, of course, very very bad. They ran all the Jews into gas chambers and asphyxiated them and did the same thing with most of the non-Jews who objected to Fascism. The Germans were very haughty and very arrogant and always treated the Rumanians like completely second class or third class citizens.

In the American press you frequently see that the Russians came to "liberate" the Rumanians and we are all supposed to laugh at the Russians "liberating" anybody. Believe me, you don't laugh about it in Rumania. The Russians don't treat anyone like second class citizens. They may shoot a few people who consider themselves first class citizens—or a little bit above first class citizens—but the attitude of the Rumanian is that the Russians did liberate them from the Germans, and they are quite happy to see some Russian troops collect in Rumania for fear some Rumanian Fascists who may have escaped will again seize the country and again invite some Germans back. A far different picture than I anticipated that we would find and incidentally, this same general type of thing applies in Hungary and surprisingly enough in the Ukraine itself.

I did not know it at the time or I had forgotten it but the Ukraine people actually revolted against Russia and welcomed the Germans into the Ukraine. The result was that the Germans went eastward through the Ukraine like a hot sun goes through a snowdrift. They got as far as Stalingrad with practically no difficulty. But the Germans were so haughty and so arrogant that the Ukrainians decided that bad as the Russians were, the Germans were infinitely worse, so after they had welcomed the Germans on their way east, they started blowing up the bridges and railroads and sabotaging the whole German army's transportation system and left the German army pretty well cut off from supplies about 1,000 miles from home and as a result, the Germans had to surrender. The bitterness toward the Germans is very extreme in Hungary, Rumania, and the Ukraine.

To get back to the Rumanians, they showed us absolutely everything you could want to see. They took us through their tractor factory at Breslau—they took us up in the Transylvanian Alps and showed us the truly magnificent hotels that are being built in gorgeous sites as rest camps for the heroes of production—heroes out of their tractor factories or their truck factories or from their state-owned agriculture or from their state-owned newspapers or state-owned railroad, or or or.

They took us to Constanta on the Black Sea, which is their major port and a good one. And there again, they are building great hotels as rest places for the heroes of production. They took us duck hunting in the middle of the Danube Delta via private steamboat that is state-owned but private for our party and then sent a special plane down from Bucharest to take us back to the capital. We saw either an opera or a ballet or were entertained at a big dinner every evening we spent in Bucharest.

We met with the whole Rumanian cabinet for four hours and discussed in detail what they might do to improve their agriculture, which they recognized to be very bad. Actually, their yields are not too bad. They have cross-bred corn that is at least as good as the old open-pollinated we used to have. They have good-looking yields of cabbage, of potatoes, of corn, of vine crops, and they had good looking wheat stubble and excellent green winter wheat up to fine stands and looking fine. But their lack of mechanization is simply beyond description.

They plant three times as many kernels of corn as they expect to harvest of mature stock and then thin by hand. They use oxen and horses for the bulk of their power. With 17 million people in their country, 11 million of them are on farms and the 11 million people on farms only have twenty thousand tractors. More than half of their wagon boxes are made out of woven willows—wicker wagon boxes from which they cannot scoop so the bulk of the corn is even picked out of the wagon box by hand. Most of the wells are open and they water the cattle by pulling it out three gallons at a time. They have the most primitive of equipment. But they are happy and proud because they are making progress everywhere. Twenty thousand tractors for 11 million people is not very many tractors. But five years ago they didn't have any tractors. And at every collectivist farm they would give us a report on the progress of that farm and it was always one of progress. They had more hogs, more cattle, more horses, were getting more eggs per hen, more milk per cow, bigger yields, better living conditions—they were in every way better off than they were five years ago.

They rather generally confessed that ten years ago practically none of them could read or write. They have had a very heavy campaign on removing illiteracy and most of them can now at least read a little and write a little. They are getting electricity on some of the farms—they have a few trucks everywhere, a few automobiles—and they are proud of their achievement and very happy with their progress. Their present situation is pretty terrible by our standards but infinitely better than their past.

At the Cabinet meeting we recommended the purchase of [a] substantial quantity of American hybrid corn for planting in 1956 to demonstrate its benefits, and they finally agreed to buy 1000 tons. Then we recommended the purchase of a minimum of ten complete sets of the most modern American corn production machinery from start to finish—tractors, discs, rotary hoes, planters with fertilizer attachments, cultivators, pickers, new modern dump wagons, elevators, stalk choppers, potato planters, potato diggers, spray rigs, insecticides, herbicides, fertilizer—just everything that we use on our own farms here at Coon Rapids so that they could have one hundred-hectare demonstrations of the most modern machinery on ten different areas within their own country. This suggestion they welcomed and are going to act upon.

So we invited three of them to come over to conclude the purchase of the corn and buy the necessary machinery.

Hungary was almost a repeat performance of Rumania. They did not definitely order any corn but they did say that they would authorize the three men they are sending over to make the purchase of corn here and to buy some hybrid seed corn processing equipment here and to buy at least two or three sets of farm machinery here.

But in the process of inviting the Rumanians and the Hungarians we ran into a rather astonishing situation. The American legation at Bucharest and at Budapest each told us that it would be a slow job to get visas for Rumanians or Hungarians away from our own State Department. This left me completely stumped. But you see, we were the first two Americans to get any kind of liberties in either country in nearly twenty years. The State Department in Washington validated our passport for both Rumania and Hungary but didn't think those countries would issue us visas. They were sure we wouldn't be permitted to see anything and doubted that we could even get into the countries. Our legations in both Budapest and Bucharest were completely astonished at the utter friendliness with which we were received everywhere.

I pointed out to them that the president, Secretary of State [John Foster] Dulles, every congressman and senator, and the great bulk of the American people had expressed greater concern about establishing friendship with the Danube Valley countries than they had with Russia itself and that I thought we had done a wonderful job of breaking the ice and it was unthinkable—and utterly stupid. I told them that I didn't want to argue about it and didn't intend to argue but, of course, I did intend to come home and tell the truth and that I had been invited to write articles not only for the Associated Press in the form of interviews but articles for several national magazines of great prominence. I told them I would hate to have to entitle my writings "Whose Iron Curtain?" and would, of course, not do so unless forced to do so by the attitude they expressed, which I simply could not believe would prove to [be] the attitude of the State Department in Washington.

But they worried me enough about the situation so that Geza Schutz and I decided to split up. He left Budapest a day or two after I did for Geneva because we had not quite finished with the cabinet when I left. I came back and saw the American State Department and he flew over to Geneva and visited with Dulles. He will arrive in the United States this Friday but I haven't been able to talk to him or get a cable from him since.

I did notice, however, that Mr. Dulles made a pronouncement before the world last Monday at the foreign ministers' conference that he proposed rolling the iron curtain clear up and having the greatest kind of ease in communications both ways and in people both ways, which I am sure is the correct attitude.

So to conclude, at least for this round, this is the way the situation stands at the moment. The Russians have definitely ordered, subject to their inspection, of course, a minimum of forty-five hundred or five thousand tons of hybrid corn as per the Matskevich letter, which I wrote on September 17, and they will want it delivered at New Orleans by late January to be paid in dollars. The Rumanians have ordered one thousand tons of corn on the same basis. The Hungarians have not specifically ordered any amount, but I would guess them at a minimum of five hundred tons [and] a maximum of one thousand tons. Delegations have been invited from each country. At the start these delegations were discouraged by the legations in their respective countries from applying for visas by warning them that the visas might be rejected so they did not actually apply for fear of rejection.

On my return to Washington, I went to the State Department itself with some pretty strong language and told them that I didn't want to have to write an article entitled "Whose Iron Curtain?" and I thought that was a pretty effective attitude. They promised immediate and serious considerations to any applications for visas, and I went immediately to the Russian, Hungarian, and Rumanian legations in Washington and got their governments to make the visas—or I think I did so—that is, the visa application. It takes about a week to get any communication, but I now have confidence that the delegations will arrive within the next couple of weeks.

Long as this letter is, it is only a part of what I would like to tell you all and will tell you all when the opportunity presents itself.

Sincerely,

Roswell Garst

Rumania

From Virgil Gligor,
Silviu Brucan, Grigor Obrejanu[1]

Dear Mr. Garst,

We were delighted with your letter. In it we found the friendly spirit which you have shown our country from the very beginning, your sincere desire to share with us your rich experience and to make the demonstration of hybrid seed corn in Rumania a full success.

Your letter brought back to our minds the wonderful days we spent at Coon Rapids with your family and friends. As to our thoughts and feelings arising out of our trip to the United States, we have expressed them publicly in our letter addressed to the main American papers, and which you have appreciated so favourably.

We wish to tell you that as soon as we returned home, we informed our government fully about our visit to the U.S.A. The warm and hospitable way in which we were received by you and other farmers of Iowa, as well as by your wise brother, Jonathan Garst, in California, made a deep impression.

We, in Rumania, highly appreciate the outspoken and courageous stand you have taken in problems concerning the improvement of relations between the U.S.A. and R.P.R., commercial exchanges and mutual visits.

We note that the hybrid seed corn has been dispatched in good condition and wish to assure you that efficient measures have been taken for transshipment at Istanbul.

As far as we are concerned, we have done everything that was necessary so that the Rumanian import-export organizations fulfill in time all the terms provided under contract, both in respect to the corn and the machines and chemicals.

---

1. Members of the first Rumanian delegation to Coon Rapids in 1955. Gligor was deputy minister of agriculture, Brucan a newspaper editor and member of Parliament, Obrejanu a leading plant geneticist.

You will have noticed that the Letter of Credit for the corn was immediately set up. We hope you will appreciate the fact that a Letter of Credit has also been set up for the machines, although in the contract this was conditioned by export license.

Your arguments contained in the letter you sent us before we left the U.S.A. recommending the reinstitution of the third and fourth schedule have been fully endorsed by our Ministry of Agriculture. We have cabled Mr. [Theodore] Huc[2] in the middle of January asking him to order nine spreaders, Lundell choppers, elevators, wagons as well as the chemicals (Aldrin, D.D.T., Dieldrin, herbicides and fungicides). We were surprised to hear that you had not been informed of this decision by January 21. Probably Mr. Huc was late in doing so. In the meantime, you will have heard from Mr. Huc, who has informed us by wire from Chicago on January 25 that he has communicated with you.

We fully share your hope . . . that the demonstration of hybrid seed corn in Rumania will prove a big success. We have chosen the fields that will be sown to this corn, the mechanics who will operate the machines, the place where Mr. [Harold] Smouse[3] will instruct the mechanics, as well as the men who are going to conduct this important experiment. We have also selected the interpreters for Mr. Smouse, Steve and Dave.[4]

The main problem now for us is to have the machines in time—that is, by the end of March at the latest. As you will realize, we are thinking in the first place of sowing machines and fertilizer spreaders which are necessary for the spring campaign. This means that the term you mentioned for shipment of these machines from New Orleans, February 15–20 should not be overstepped. Knowing you as we do, we feel certain that this target will be reached.

Our Ministry of Agriculture has been very favourably impressed by the fact that you are sending Mr. Smouse and your clever sons, Steve and Dave, to Rumania in order to make the demonstration of hybrid seed corn a 100 percent success. Be certain that they and their wives will have the best conditions. We shall assist them in every way not only on the job but also to make their stay here as pleasant as we possibly can. We still hope that you and Mrs. Garst will find the time to visit Rumania this summer perhaps after your planned trip to the Soviet Union. We do think you should come

2. Agricultural attaché. Rumanian Legation, Washington, D.C.
3. A mechanic for Grettenberg Implement Company of Coon Rapids, who accompanied David Garst and stayed on during Stephen's visit to keep the machines in working order and to train Rumanian mechanics.
4. Garst's sons, who farmed in Coon Rapids.

since during your short stay in Rumania you have been able to see only a small part of our picturesque beauty spots. And now that you have successfully passed your exams in hunting wild geese and ducks, we think of organizing a bear-hunt for you in the Carpathians. . . .

We suppose you will have to receive your wives' permission for this hunting-party, which may prove a more difficult matter than to obtain an export license from the State Department.

Now back to business. As you know, our government has asked for visas for two Rumanian commercial representatives who will negotiate the exchange of cement for tallow, cotton, hides, etc. Both are experts in their respective fields and experienced negotiators. We hope that their visas will arrive without much delay.

Messrs. [Joseph] Hincu[5] and Csendes[6] also have powers to negotiate other business with U.S. firms. We too hope that following President Eisenhower's message to Congress,[7] your government will have a more favourable attitude towards trade between East and West.

There are plenty of opportunities for commercial exchanges. A few days ago an American businessman who visited Rumania bought window glass from us worth $200,000. Messrs. Hincu and Csendes will bring samples and printed matter concerning a wide range of Rumanian goods that may be of interest in the U.S.: cement, furniture, window glass, games hides, down, fine feathers, national costume, embroideries and needle-work, caviare, Sibui salami, wine, etc. which we could exchange against American merchandise.

Our government intends to open a permanent trade mission in New York soon, and to send an agricultural attaché. As you will remember, we have discussed this with you and with officials of your State and Agriculture Departments during our visit to Washington. . . .

We have not forgotten our promises either. Messrs. Hincu and Csendes will bring you Rumanian wines and Sibiu salami. Mr. Brucan will procure a good copy of "Jew with a Goose" by the Rumanian painter [Nicolae] Grigorescu[8] which you liked so much. We shall also send you some more snaps taken during your visit to Rumania.

5. Economic counselor, Rumanian Legation.
6. We have been unable to obtain further information on Csendes.
7. In his third annual State of the Union Message, 5 January 1956, Eisenhower talked about expanding trade and encouraging economic growth and stability as ways of providing peace and prosperity and at the same time reducing international tensions.
8. Nicolae Grigorescu (1838–1907) usually painted aspects of Rumanian peasant life: picturesque villages, old country roads, and the characteristic beauty of his native countryside.

In conclusion, we wish to tell you that we wholeheartedly share your hopes concerning the development of relations between our two countries and that we are looking forward to the abolition of the obstacles still standing in the way. . . .

Yours very sincerely,
Virgil Gligor
Silviu Brucan
Grigor Obrejanu

May 22, 1956

To Mr. Silviu Brucan, Minister Plenipotentiary
The Rumanian Legation, Washington, D.C.

Dear Minister Brucan:

Following our weekend discussions, I wish to outline my thoughts in such form that they can be transmitted by you to Bucharest for study by interested people there, such as the Minister of Agriculture and Deputy Minister Gligor, and I hope perhaps the President of your Academy of Science,[1] Mr. [Traian] Savulescu, and his wife.

The hybrid seed corn and the farm machinery that are being used in your country this year are simply the first steps in improved agriculture. They are important first steps—but only first steps.

Really high corn yields with low labor require hybrid seed and mechanization, but they require two other things—fertilization and, finally, irrigation. Hybrid seed plus mechanization plus fertilization plus irrigation simply has to be your final goal. Perhaps the proper use of insecticides and herbicides should be added, but they will come naturally.

Those four steps can more than double corn yields on every field where they are used above anything you have been in the habit of raising—and you can do it with as little as 10 percent of the man-hours you have been using.

I think learning the complete picture of high corn yields with low man-hours should be our next year's objective. That is, I think heavy fertilization and good irrigation should be added to at least a few fields along with the hybrid seed and mechanization that you are using this year.

The next step of course is to learn the most efficient way to use the

---

1. Known as the Academy of the Rumanian People's Rebublic.

corn after it has been raised—the most efficient ways to transfer it into meat and eggs and milk and fats and hides. I wish to give you only the briefest kind of an outline of what we did discuss along these lines and what I think would be good for your country to start in with.

First, we want to discuss chickens—both chickens for eggs and chickens for meat. Chickens for meat in this country are butchered when they are ten weeks old weighing about three pounds and are called "broilers." You have been able to buy them at the grocery store and they are truly delicious, and very fast, and they require relatively low capital investment.

Ten or 15 years ago it used to take five pounds of feed per pound of broiler meat. A few years later it took four pounds of feed per pound of meat, a few years ago it took three pounds of feed per pound of meat, and now the standard is 2.5 pounds of feed per pound of meat. What we now do in ten weeks used to take sixteen weeks.

A similar increase in the efficiencies of the production of eggs has occurred in the last ten years.

Furthermore, we have made rather comparable progress in the efficiency with which we raise hogs. It was the knowledge of how to use antibiotics and vitamins and minerals in proper amounts, along with the grains, that let us make such a phenomenal increase in the efficiency of hogs and chickens.

At the same time that these discoveries were being made on increasing the efficiency of hogs and chickens, some equally effective discoveries were made on how to get real efficiency out of feeding celluloses, such as corn cobs and cornstalks, to ruminants, such as cattle, sheep, and goats. We discovered that when fed with ample proteins these celluloses were highly successful. Then we discovered that the urea form of protein worked wonderfully well as protein for ruminants, if the urea was accompanied by plenty of minerals and by a fast carbohydrate such as molasses.

I telephoned to Mr. [Philip] Maguire[2] this morning and told him that I felt it was imperative that your agricultural attaché arrived by June 1 and if you have any trouble at all in obtaining the visa promptly from our State Department, I want you to get in touch with Mr. Maguire because I believe that he can obtain a prompt issuance of the visa. If not, he can report to me and I will try to accelerate the issuance.

As you can see from this outline, your agricultural attaché will have much to look at and to study, and to take pictures of, and to get full

---

2. Garst's Washington attorney. They met during New Deal days, when Maguire worked for Milo Perkins in the USDA.

information about. I am sure that Mr. [Drew] Pearson[3] will be an excellent guide and driver and interpreter. I am in a position to see that they get to talk to the practical farmers who are doing all the things outlined above— and study the most advanced methods.

After your agricultural attaché has pictured and studied and reported upon to your government in Bucharest all of the items I have suggested in this letter, plans can be made to put on demonstrations in Rumania next year covering all of the different items. . . .

I am completely certain that the wise thing to do is learn everything possible here first—and then put on the best possible demonstration a year later in your own country. . . .

Stephen suggested that I wait in writing this letter until after David's return,[4] but time flies on, and I wanted your government to know my thoughts in this matter at the earliest possible moment, and I again wanted to emphasize to you the desirability of your agricultural attaché's arrival by June 1 if that is possible.

> Very sincerely yours,
> Roswell Garst

November 8, 1957

To Mr. V.J. Tereshtenko[1]
219 E. 5th Street
New York 3, N.Y.

Dear Valery,

I think you must be going crazy waiting for something to happen.

Well, anyhow, this is the story. The Canadian government invited the Rumanians to send a farm delegation to Canada. They placed an order with us from Washington, D.C., for fifteen hundred tons which did not match our supply—that is, they wanted more hybrids than we could supply in some varieties and not as much as we could supply in other varieties. So I decided I would have to go to Rumania and so told Minister Brucan.

---

3. The son of Leon Pearson, an NBC executive, and a nephew of the Washington commentator Drew Pearson.
4. David Garst was in Rumania at this time, supervising the planting of hybrid corn. Stephen would take his place in the summer of 1956 and stay through the harvest.
1. First hired from Columbia University to interpret for the Russian delegation of 1955, Valery Tereshtenko later became a commercial agent and friend of Garst.

He replied that the delegation was now in Canada so I flew to Ottawa and visited with them—found out how much they needed in the way of single crosses and varieties—and came home and am going back to Winnipeg, Canada, to see them tomorrow. I am sure they will buy fifteen hundred tons or more, and I think they might buy as much as three thousand tons.

Now, when I thought I had to go over to see them, I put on a good tough struggle with the State Department to try to get my passport validated for Bulgaria. They absolutely refused to let me go into Bulgaria but will issue an export license if I sell the Bulgarians corn while I am in Bucharest.

I think I weakened them a good deal, and I think they are willing now to almost give me the passport but not quite. I think if I could talk to the top people in the Bulgarian government—not only their minister of agriculture but perhaps their minister of foreign affairs—while I am in Bucharest, that I could come home with twice as much power to use on the State Department as I had when I went.

Mr. [Romuald] Spasowski, the ambassador from Poland, is supposed to look after the Bulgarians' interest in Washington. But the way they do it is very slow and roundabout. He wires Warsaw and tells his government, then his government tells the Bulgarian representative in Warsaw, who then wires Sofia, and then Sofia wires the Polish Embassy at Warsaw, who then informs Mr. Spasowski. If everybody were home—which they won't be; it would take thirty days—even if they didn't have the Asiatic flu, which they might have.

I don't know why I neglected to have you see [Peter] Voutov, the Bulgarian representative at the United Nations, whom I found so delightful when I met him and whom I was so anxious to have visit Coon Rapids.

I wonder if he would be willing to cable his government that I expect to be in Bucharest and recommend to them that the minister of agriculture and some high official of the Foreign Affairs Department meet with me there, telling them that I can take their order for corn, that I have been assured I can obtain an export license for it, and that I would like very much to discuss the whole situation with them because I think I might be able to have some effect on the State Department when I return. . . .

I'll try to call you tonight and discuss it by phone. Don't show this letter to anyone. Just go down and see Voutov and find out if he can make arrangements for me to meet at least their minister of agriculture and somebody from their Foreign Affairs Department in Bucharest and if they are willing to place an order for the corn. They wanted five hundred tons last year; with the Rumanians' buying and raising twice as much corn as they had last year I would think that the Bulgarians might well want one thousand tons.

With two satellites whirling around the earth, the Soviet Union can brag, of course, on their technical ability. It seems to me that the offset is for us not to be bragging too much about our technical ability on satellites— but to brag on our technical ability in farming, in mass manufacturing, in plumbing, in electricity, in roads. We are so far ahead in those fields that I think we ought to brag on it, and then I think we ought to show them.

Anyhow, I'll try to call you.

Sincerely,
Roswell Garst

January 16, 1960
To Vice-President Alexandru Moghioros,
The Rumanian People's Republic,
Bucharest

Dear Vice President Moghioros,

In the fall of 1956 when Mrs. Garst, Mr. and Mrs. [John] Mathys,[1] and Dr. William Brown[2] and I visited in Rumania, it was the unanimous opinion of each of us that it would be wonderful to have you visit the United States. You will probably remember that I discussed the matter with you on the train when we went down to look at the hybrid seed corn field trials in Eastern Rumania. . . .

I think the matter of protocol may have raised some questions, at least in the mind of Minister [of Foreign Affairs] Macovescu and Ambassador Brucan and perhaps in your own mind—and in the minds of your associates. You may remember that I told you on the train in the fall of 1956 that Mrs. Garst and the Mathys' and Dr. Brown all raised the question of protocol when I had told them that I intended to invite you. They explained carefully to me that as an American citizen, I was not entitled to invite a vice president of Rumania, that the invitation had to be issued by the United States government. I never did believe that that was true. I do not believe it

---

1. Head of the Garden and Packet Division of Northup, King and Company of Minneapolis.
2. At that time, chief geneticist for Pioneer Hi-Bred International.

yet, and I have discussed the matter at length with the State Department of the United States government. The State Department of the United States government is perfectly willing to have me invite you to the United States as my personal guest to show you all the parts of American agriculture which I feel so strongly can be helpful to your country in the future. The State Department recognizes that an unofficial visit as my personal guest would have great advantages to you in that it would not entail a very great amount of publicity with too much pressure from the American press, American radio, American television, and other news media. You could see four times as much, learn ten times as much, travel with much greater convenience—and I am sure the trip would be infinitely easier and more effective than it would be if you came as an official guest of the American government.

In fact, I would like to invite not only you, but I would like at the same time to invite Mr. [Mihail] Florescu, minister of petroleum and chemicals. And I would really prefer if only the two of you, with a really good interpreter, would come to the United States—the last week in April, I think, would be an ideal time—and stay for not too long a period—say a period of perhaps two weeks. I would be happy to meet you in New York on your arrival, stop briefly in Washington, come out to Iowa and spend several days in Iowa, take you out to the irrigated section of Nebraska, take you on out to California, and show you the delta area of the San Joaquin and Sacramento rivers, which is so much like the delta area of the Danube, and show you the broad implications of American agriculture that can be duplicated in Rumania.

The 1960s can see very, very terrific increases in agricultural production in Rumania. The techniques that we are using can be applied perfectly in Rumania. Your production can not only be doubled—in my opinion it can be tripled in a ten-year period. And specifically, the reason I want Mr. Florescu to accompany you is that it is a combination of chemical knowhow and agricultural knowhow that accomplishes this terrific explosion in agricultural production that is so probable. You and Mr. Florescu together can accomplish production increases that sound impossible to people who have not seen and partaken in the American agricultural production explosion.

Your country has already made great progress agriculturally. You have changed from open-pollinated corn to hybrid corn. That was the first big and necessary step. You have made a good start—but only a good start—on mechanization. You have seen how it is possible to reduce the man-hours required to grow a bushel of corn very greatly with hybrid corn and mechanization. Those are two great steps forward that you are already taking.

You need to see the full development of American irrigation, and you need to see that full development of irrigation in combination with the full development of the chemicals of agriculture. From a volume standpoint, of course, the chemical fixation of nitrogen is of first importance: more nitrogen fertilizer, more complete fertilizers containing phosphorus and potash as well as nitrogen, fertilizers that are of good physical condition so that they are convenient to use, fertilizers that are economically produced. But in the United States, we are now going beyond just the chemicals in fertilizers. Chemical soil insecticides are very great tools for getting better stands of beets, corn, and other crops. And now we are even adding chemical herbicides to help us control weeds. It may interest you to know that a full third of the total production of the agriculture of the United States is due to chemicals—chemical fertilizers, chemical insecticides, chemical herbicides. And it may further astonish you to know that in the case of the Garst family operation, we attribute a full half of our total production to the heavy use of chemicals in all forms. We of the Garst family are producing twice as much in the way of food as we could be producing without the use of chemicals.

In those drier parts of Rumania where lack of natural precipitation handicaps crop production, irrigation is a simple must. You must go to irrigation. And there are several ways to irrigate—all of which probably have a place, all of which are practiced in the United States and rather well perfected. Pump irrigation is widespread in Nebraska. And gravity irrigation of streams is also practiced widely in Nebraska and in Colorado. All of these things I would love to show you.

I am particularly intrigued because your country is rich in both oil and natural gas, and the whole basis of our chemistry is this use of oil and natural gas in the production of proper chemicals that permit the very great explosion in agriculture that I am sure is bound to come. You are rich in natural resources, you are rich in the desire of your country for greater production, you are rich in the intelligence of your people and their education. . . .

Not only will Mrs. Garst and I feel honored by such a visit—both Stephen and David are very anxious to have you come and very anxious to be helpful to both you and Mr. Florescu—and through you, helpful to your country. Our associations with your country have always been most pleasant, and we think this is the most helpful thing we could do for the agriculture of Rumania. . . .

Very respectfully yours,
Roswell Garst

November 29, 1960

To Mr. Bucor Schiopu, Minister of State Farms
Bucharest, Rumania

Dear Minister Bucor Schiopu,

John Chrystal[1] returned in September from a thoroughly delightful trip—a trip of very high interest—not only to Rumania but previously to the Soviet Union, Hungary, and Bulgaria. He got home just in time for harvest and we have all been so extremely busy this fall that it was not until today that Mr. Chrystal and I write you this very much delayed letter of appreciation and congratulation. Naturally, Stephen and David and John Chrystal and I have talked a great deal about Rumania and the progress that has been made there since I first visited your country in 1955.

We almost think we should take out citizenship papers in Rumania because visits to your country have occurred every year by some one of us since 1955. So we have been able, amongst hands, to keep track of your agricultural progress on more or less an annual basis. Mr. Chrystal is sitting with me as I dictate this letter and contributing greatly to it.

What progress you have made! Mr. Chrystal is specific in saying that Fundulea[2] is the best corn research establishment in Eastern Europe. And also that you have some extremely interesting grain sorghum plots there.

You must have great pride—justified pride—in the progress you are making. Between 1955 and 1960, you have gone from no use of hybrid seed corn to nearly complete use of hybrid corn. It took the United States fifteen years to go from no hybrid corn to practically complete hybrid corn. This, it seems to all of us, is as it should be. You are taking advantage of the mistakes we made and dividing the time requirement by about three.

When I was there in 1955, each one of the state farms and collectivist farms bragged about how many more horses they had in 1955 than they had had in 1950. When I was there in 1959, you told me yourself that you now had thirty-five thousand fewer horses than you had previously had because you were going to tractors so rapidly. This again is as it should be—and now horses are going to be animals of pleasure with you shortly, as they are now in the United States. So both Mr. Chrystal and I congratulate you and your associates highly for the very great progress you have been making. It probably seems slow to you because you see it every day. It seems remarkable to us because we see it once a year. . . .

1. Garst's nephew, who in June 1960 took Garst's place as a guest of the Soviet government. It was Chrystal's first trip to Russia and Eastern Europe.
2. Encouraged by Garst, this principal Rumanian agricultural research institute greatly expanded its hybrid research operations, beginning in 1955. It now coordinates research at thirteen other institutes, six central stations, and forty-one experiment stations.

We wish to make several specific recommendations as to several more steps that you should take. In the interest of brevity and being specific, I will list them in order we think are most important—probably more important timewise than ultimately. They are as follows:

1. We wish to recommend the purchase immediately of enough hybrid grain sorghum seeds to get a very widespread use of hybrid grain sorghums in Rumania in the year 1961. We are bold enough to suggest the purchase of five thousand fifty-pound bags. It should be planted at the rate of four hectares per bag of fifty pounds—and we think you should not start with less than twenty thousand hectares very widely distributed over Rumania. Garst and Thomas could supply you with that amount at a price of twelve cents per pound—six dollars a bag—of seed of very high germination and great seedling vigor, treated with both a fungicide and an insecticide and double-bagged for foreign shipment F.O.B. the American port.

The hybrid grain sorghums have given phenomenal success in the drier areas of western Nebraska, western Kansas, and eastern Colorado and have absolutely revolutionized the cropping system of that area. Before the time of hybrid grain sorghums, the United States used to raise about 250 million bushels with an average yield of just over 20 bushels per acre. This year, the United States produced more than 600 million bushels of grain sorghums with an average yield of 40 bushels per acre. This has all been accomplished in the last five years or thereabouts. We believe this should be ordered immediately—for the earliest possible shipment. Also, Garst and Thomas will be willing to supply you with any hybrid seed corn that you may wish to purchase. Would you kindly let us know your desires both as to hybrid seed corn and as to hybrid grain sorghums?

2. We are completely certain that you should purchase very substantial quantities of foundation stock for egg laying chickens from Pioneer Hy-Line Company of Johnston, Iowa, and embark upon a very large extension and a very great improvement in your egg laying facilities—and at the same time, you should buy some of the foundation stock for the meat type of chickens for the production of broilers. An expansion in your whole poultry industry is one of your most important steps to take.

In this connection, we are extremely happy to know that George Finley, Jr.,[3] is now in Rumania, putting up the universal

3. Son of the president of Finco, the Illinois-based company that supplied Rumania and the Soviet Union with processing plants.

building which can contribute greatly to the mixing of feed for all types of livestock—not only chickens but pigs and cattle as well. Also in connection with chickens, we think it is essential that you buy mechanical equipment for hatching, feeding, housing, butchering, refrigerating, and transporting. Mr. George Finley and the Finco Company have just finished the shipment of this type of equipment to the Soviet Union and I am certain should serve you well in helping you plan the full chicken setup not only for egg layers but for broilers. You have to have three broad things with chickens— excellent blood lines (the genetic characteristics), absolutely proper feeding equipment and proper feeds, and finally, the best of housing and other mechanical aids. With all three, chickens are the most efficient method of transferring coarse grains into delightful and nutritious human food.

3. We think you should definitely proceed with some irrigation. You have the whole vast plain of highly fertile land that is too dry for maximum production. You have water available—and you could triple and multiply by four your crop yields. We think you should budget a substantial amount of money to get started in this direction.

4. We are definitely sure that road grading equipment is one of the biggest steps forward that could possibly be taken in Rumania. Nothing is more phenomenal than the progress that has been made in road grading equipment in the United States. It requires very large crawler-type tractors with bulldozers and large automatic scoops and hydraulic road grading equipment, but with the proper equipment, a relatively few men can grade mile after mile of road and the men do not have to be highly skilled—any good mechanic who knows how to service diesel tractors can do a phenomenal job. As your agriculture improves, your country road system must improve so that you can deliver to market—and deliver to the principal highways and to the principal cities and larger towns—the produce of the farm. I can sincerely tell you that in my lifetime I think the progress in economical roadbuilding in the United States has made the greatest progress. So we highly recommend at least a start in this direction. The machines are extremely expensive—but effective beyond the wildest imagination. Three men can grade a mile of road reasonably well in a day's time—and in the course of a summer, three sets of such machinery could grade a phenomenal quantity of road.

In the United States, we grade the roads up several feet higher than the surrounding area through which they pass so that the snow blows off from them and they are all-weather roads then, winter or summer—which is of great importance. We are sure you should get started immediately in this direction.

We both feel that you yourself—or one of your trusted deputies—should come to the United States at the earliest possible moment and give this group of things investigation for immediate purchase. A small delegation fully authorized to make the purchases could accomplish wonders in a very short time. Actually you could order whatever hybrid grain sorghum seed or hybrid seed corn you want by mail or by cable. However, on such things as road grading equipment and buildings for chickens, you should have a representative here. . . .

Mr. Chrystal and I are both very apologetic for the long delay before we got around to writing this joint letter. I can only tell you that we have just finished the largest harvest in all history—and that we have been extraordinarily busy. Garst and Thomas raised the finest seed crop in all of history of both corn and grain sorghums—and both David and I have been working at the sales of hybrid corn in the United States.

Stephen had twenty-five hundred acres of corn this year on the Garst farming operation that had an average yield of above one hundred bushels per acre. We used heavy fertilization, we used insecticides, and we tried a new herbicide, Atrazine, made by the Geigy Company of Switzerland, which we found to be extremely effective.

We told Steve and Dave this morning that we were writing this joint letter, and they both asked us to include their best regards to you and to their many friends in Rumania along with our own.

We will appreciate hearing from you.

> Very sincerely yours,
> Roswell Garst
> John Chrystal

May 17, 1961

To Ambassador Silviu Brucan
and Madame [Sasha] Brucan
New York

Dear Ambassador and Mrs. Brucan,

. . . A peculiar thing about the times is that I feel somewhat more optimistic about the international situation than I have in the past. More and more people are coming to the realization that with hydrogen and atomic bombs, war is completely unthinkable. At the same time, everybody is beginning to realize that we are going to have well over five billion people on the face of the earth before we reach the next century. I suspect actually we will pass five billion people by 1990.

I think you will be interested in the broad direction that I hope the international political situation will take. Your Communist area has been scornfully calling us Capitalistic Imperialists. We have been calling you Communists who are trying to take over the whole world. So the Communist bloc countries and the Western countries have been quarreling and hurling accusations at each other and getting exactly nowhere.

A division ought to be made along different lines!

As I see it, there are two classes of people in the world—those who can read and write and those who can't read and write. We'd better divide up along that line. The Communists can read and write. So can Western Europe and the United States and Canada and a few other isolated spots on the face of the globe, but something less than half of the people in the world can read and write.

Actually, the people who can read and write are quite rapidly getting a better standard of living whether they be communists or noncommunists. The people who can't read and write are getting hungrier and hungrier— and it's going to be more and more difficult to help raise their standards of living.

The people who can read and write are going to become the "have" nations—whether they be Communists or not. The people who cannot read and write are going to continue to be the "have not" nations.

I am beginning to become hopeful that the countries that can read and write will find it to their own advantage to quit quarreling about Communism as opposed to what you are inclined to call Capitalist Imperialism—I am inclined to think that the nations who are literate better get together and start helping the nations who are illiterate or it will continue to be a pretty sad world in which we live. . . .

Sincerely,
Roswell Garst

May 26, 1961

From Silviu Brucan
New York

Dear Mr. Garst,

I was very impressed with your letter and although diplomats are supposed to have no personal feelings—just governmental instructions, I want you to know how much I appreciate your thoughtfulness and courtesy.

As a matter of fact, I consider the days I have spent in Coon Rapids among the brightest and the most pleasant during my stay in the States. I always remember with joy, pride and hope our conversations and I have no hesitation to state that I learned a lot from you not only about hybrid corn, but about man and life as well.

I found your new world-division along literate and illiterate lines most interesting. It is a striking fresh approach to get the literate bulls together!

I would add however, an amendment to your solution to world problems. What if literate nations think of those problems not in emotional terms but rather in rational terms?

We find everyday in the press, on radio or T.V., positions on international issues described as "tough" versus "soft", and "rigid" and "inflexible", and the alternative for negotiations is between "standing firm" and "standing pat"! This is the emotional approach, which is usually the childish or in the best case the teen-age way of reacting to events.

Don't you feel that it is about time for literate nations to switch from childhood to adulthood in international affairs, i.e. from the emotional approach to the rational approach and to seek not what is "tough" or "soft", but what is intelligent, sane, reasonable?

I should say that if we all literate nations behave like thinking beings, the world will be in a much better shape. If, however, emotional reactions prevail, the circumstance that we are literate won't help too much in avoiding the catastrophe.

. . . My best regards to Mrs. Garst and to the whole Garst family.

Affectionately yours,
Silviu Brucan

June 5, 1963

To Mr. Gheorghe Gheorghiu-Dej
President, Council of Ministers
Peoples Republic of Rumania

Dear Sir:

After an enjoyable trip home, with brief stops in Athens and Rome, Mr. Chrystal and I returned to find crops here in Iowa looking fine.

Mr. Chrystal and my son, Stephen, are in Chicago today finding out whether corn combines and two-row self-propelled silage harvesters can be sent to Rumania in time for this year's harvest.

After a good deal of conversation on how we can be of greatest help, we have concluded that the first thing to do is to send three bulletins over to you for interpretation into Rumanian and for study. They tell the story very simply of what we have accomplished in the United States in the last few years—and how we have done it.

The bulletin "Congratulations to You" records the phenomenal increase in yields of corn and grain sorghum. Before 1958, the United States never had an average yield of corn as high as forty bushels per acre. In 1961 and 1962 the average yield was above sixty bushels per acre. That is a 50 percent increase in a five-year period. That increase was almost wholly due to more and better fertilizers, wider use of insecticides and herbicides—in fact, *more and better chemicals.*

With your petroleum and chemical ministry under Mr. [M. Mihail] Florescu, you can and should go as far and as fast as we have done and as soon as possible.

The second bulletin "Progress and Opportunities in Corn Growing" spells out how to lower labor requirements per unit of feed or food.

It requires the same labor per hectare to raise a low yielding crop as a high yielding crop. If you double the yield per hectare by the use of heavy fertilizer applications and the use of proper insecticides and herbicides, you cut the labor requirement per ton of corn or ton of silage by half.

Your tractor factory is now making fine tractors, but it is our considered judgment that when it comes to other tools, we are still ahead of your country.

We, in the United States, have made great improvements in planting equipment, in fertilizer application, in cultivators, and especially in harvesting equipment, both for grain and for silage.

We are strongly urging the purchase of several sets of corn growing machinery of the most modern American design so that you may see the improvements that have been made in the last few years.

Also, as we told you, we feel that American machinery for the building of farm to market roads is superior to that of any country because we in the United States have more farm to market roads than any similar area in the world and by a wide margin.

We believe you should buy and use several sets of road building equipment for country roads. . . .

I want you to know how proud I was to see the great progress that has been made in Rumanian agriculture since I first visited your country in 1955.

At that time, you had no hybrid corn, little mechanization, no broiler industry, few labor-saving devices. I was not critical because I knew that you had only recently thrown off the yoke of feudalism. It was my estimate in the fall of 1955 that it was taking Rumania two hours of man labor to produce a bushel of corn. Much of the plowing was being done by horses or even oxen.

The United States was then, in 1955, requiring six minutes of man time to produce a bushel of corn.

I suspect the average of the United States is now down to four or five minutes per bushel of corn. And you are down to thirty minutes per bushel. And that half of your labor requirement is the harvesting.

You must greatly reduce harvesting [time] by the use of corn pickers and/or combines. And when it comes to silage, you must use modern silage-chopping machinery, not only to lessen the labor but also to improve the quality.

You have made phenomenal progress. The broiler installation we saw was grand. You only need to multiply it in numbers. The egg-laying layout was fine. Your most modern dairy farm was as good as any I have seen in the United States.

The fact that you are building more and better fertilizer plants gives great promise of increased yields.

And finally, I am sending you a bulletin on the use of urea as a protein for your cattle and sheep.

By following this bulletin carefully, you can divert the sunflower meal and flaxseed meal from the cattle to the swine and chickens and have protein in more ample quantities for all livestock and poultry.

Mr. Chrystal and I will try to get every bit of information requested by all of those in Rumania who had questions. It will take a little time.

We did not want to delay this letter of appreciation to you and all the others who did so much to make our visit so delightful.

Respectfully yours,
Roswell Garst

Sept. 11, 1964

From Silviu Brucan
Vice-President, Ministry of Television and Radio
Bucharest

Dear Mr. Garst,

Once more, while remembering the date of my last message, you will surely notice that Rumanians are poor in writing letters, a fact which I could hardly deny.

However, you will also have to admit that I am not incorrigible and that there is still some hope I may be regained for civilized human relations.

The truth is that I am thinking very often to the unforgettable days I spent in that marvellous place called Coon Rapids, surrounded by your touching kindness and your exceedingly hospitable family. I have always shared your noble ideas on peace and friendship among nations and I must say I admired your unflagging stand in the worst days of the cold war. Last but not least, you know very well how much I appreciated your immense contribution to the modernization of our agriculture, the fair and wise advice you gave to our men in growing corn with highly productive seeds, with chemistry and modern machinery.

Well, since you called me a "farmer"—one of your favorite jokes—I feel duty bound to give a report on the most significant developments in Rumanian agriculture, although you know better than anybody else how innocent a layman I am in this respect.

Let me tell you first of all that this year 1964 *the whole corn acreage* was cultivated with double hybrid seed corn, whereas in 1960 only 20% was so cultivated. Giossan and Muresan, the two leading experts of Fundulea, were given just recently a high award for obtaining tens [sic] of indigenous varieties of hybrid seed corn adapted to the natural conditions of Rumania, early varieties as well as late ones. To give an idea of the best results obtained up to now, I will mention that as many as 70 state farms harvested in 1963 over 8000 kg of corn per acre in non-irrigated land. Surely, the yields are much lower on collective farms—and surely all of them are much lower than in Iowa. But wait and you will see.

The rhythm of mechanization is not impressive for the simple reason that we cannot speed up the process of mechanization unless we create industrial jobs for the peasants displaced by farming machinery. This is why we have to accelerate industrialization and to view the problem of economic progress both in industry and agriculture in all its intricacies. A proper balance should be maintained by all means; otherwise we get into trouble. Thus, in 1964 we have got in the fields 72.000 tractors (1 tractor per 270 acres), 62.000 mechanical planters and 34.300 grain combines, including corn pickers. It is not so bad! However, the farming equipment produced in Rumania that exceeds our needs goes to export.

I now come to fertilizers. As you know, we did not wait for a drought to start building chemical plants to produce fertilizers. The result is that this year we were in a position to supply the agriculture with 1.100.000 tons of nitrogen and phosphoric fertilizers; next year new plants will be in operation and the total output of fertilizers will reach 2.500.000–3.000.000 tons annually, which is pretty good for our arable land (3.2 million hectares wheat, 4.0 million hectares corn etc).

You will get a more comprehensive picture if I will tell you that the average grain harvest in 1960–63 was of 10 million tons as against 9 million in 1956–59. In 1963, which was a rather poor year for East European agriculture, owing to unfavourable climatic conditions, we got 10.4 million. I realize that this is not something to perplex an Iowa farmer and yet the important thing is that we are making progress.

Now that I bored you to death with plenty of figures and propaganda, let me tell you that I remain deeply grateful for all that you did to make this progress possible and I miss your friendly presence, your wisdom and witty conversation. Please convey to Mrs. Garst and to the whole Garst clan my highest regards.

Sincerely yours
Silviu Brucan

Bulgaria

April 12, 1957

To Mr. Edward L. Freers, Director
Office of Eastern European Affairs
Department of State
Washington, D.C.

Dear Mr. Freers,

Nearly a month ago, on March 22 to be exact, I wrote you at some length about my desire to invite Dr. Peter Voutov, the permanent representative and ambassador of the Bulgarian government to the United Nations, to come out to Coon Rapids for a visit and, if he wished to, to bring his wife.

I would like your permission to tell him that I have taken the matter up with you and that upon his request, permission will be granted for such traveling.

I have not as yet heard from you and I would appreciate an early reply to this request because May is one of our beautiful and active months—spring arrives, the grass is green, and it's a fine world to live in.

Incidentally, you may be interested in knowing that I have already invited Mr. Spasowski, ambassador from Poland, and Mr. Edward Iwaszkiewicz, the agricultural representative on their delegation, which is in the United States, at the invitation of the State Department negotiating for credit. They have accepted the invitation and expect to come as soon as the negotiations in Washington are completed.

And now, Freers, I am going to issue another invitation—*to you*!

I am completely serious in the invitation and think it would be very, very worthwhile for *you* to come out. I'll give you a few reasons.

You are head of the Eastern European Division. The biggest single problem of Eastern Europe is their lack of food. It's not the lack of total food—they can get their bellies full of bread, potatoes, and cabbage most of the time—it's the lack of protein and the lack of fat.

It is actually, Freers, a lack of knowledge about corn—how to raise corn and how to use corn.

I don't know whether you realize it or not, but we raise more bushels of corn in the United States than all other grains put together—and we don't eat any corn to speak of; we eat eggs and milk and beef and pork and chickens and turkeys, all of which are basically fed on corn and are the result of corn.

Of course, I am in the seed corn business and so we had to learn how to raise corn the best possible way. But then we also learned how to use all of the corn plant—we even used corn cobs for feeding cattle; we use the corn grain for not only cattle but for hogs and chickens as well. So we can do a better educational job at one point—Coon Rapids—than can be done anywhere.

I know it's a peculiar suggestion—for a man from the State Department who is in charge of Eastern European affairs to come out to a small town in the center of the Corn Belt. But I believe the information you could obtain here would be of high value to your present position there and I'm perfectly sincere in issuing the invitation and in urging you to accept it.

Very sincerely yours,
Roswell Garst

May 6, 1957

From "*Hranexport*"
Sofia, Bulgiaria

Dear Mr. Garrot,[1]

We are authorized by the Minister of Agriculture and Forests Mr. Stanko Todoroff to give you his thanks for your amability to help our agriculture to increase the gains of maize and to achieve nearer relations between our two countries.

The Ministry of Agriculture and Forests is particularly thankfull for the samples of Highbreed Maize, Sorgho and granulated Fertilizers, which you send. We inform you, that the seeds will be sown in the experimental institutes of the Ministry in three different districts of the country and separately at watering conditions for differential examination with our sorts

1. We have retained the original spelling and wording in this letter.

and crossed sorts Hybrids. We will keep you informed about the results of the examination. After the harvesting of the experiments and after we have established the results of them, we will use your amability to beg you to deliver us the Hybrid Seeds which will be most suitable for a large production experiment in 1958.

The Ministry of Agriculture and Forests agrees to receive the delivery of three complects machines for raising of the Maize, thanks you for your readiness to help in this delivery and leaves it to you to prepare and send the list and prices of these machines, taking under consideration the particularities of our country.

We agree that the delivery of the machines after your recommandation should be ordered within the firm "Grettenberg Implement Company."[2]

We beg you to receive our highest esteem and greetings, and remain, dear Sir.

Yours faithfully
"HRANEXPORT"
State Commercial
Enterprise

November 14, 1957

To Stanko Todorov
Minister of Agriculture
Peoples Socialist Republic of Bulgaria
Sofia, Bulgaria

Dear Mr. Minister:

A year ago, when I visited in Bucharest, I was informed that the commercial attaché of Bulgaria in Bucharest wished to see me, and I did call on him. He informed me that the Bulgarian government wishes to purchase five hundred tons of corn and five "sets" of American farm machinery, which was just exactly half of what the Rumanian government had purchased the year before. I told him that I would be happy to be of assistance—that Garst & Thomas could supply the seed corn and that Mr. Grettenberg could supply the machinery—but that we would have to see

2. Company located in Coon Rapids, Iowa.

about getting an export license. Because of a lack of diplomatic relationship between your country and my own, it was not absolutely certain that the export license would be available.

Upon my return to Washington I did find out that the State Department would not in any way oppose the issuance of an export license by the Department of Commerce. However, they would not grant me a validation of my passport to come directly to Bulgaria.

I have urged [upon] our own State Department the advisability of the reestablishment of actual diplomatic relationships upon you and through you upon the Bulgarian government. Speaking not as an official, but as a private citizen, I expect to continue to urge both the Bulgarian government and the government of the United States to reestablish active diplomatic relationships because I think otherwise the citizens—the people—of both countries suffer from the lack of contact. I feel this matter very strongly. And, I have no doubt that there are many other Americans—and many Bulgarian people—who would like very much to see this accomplished.

I have recently asked our State Department and the Commerce Department if they are in any way opposed to granting an export license from the United States to Bulgaria, and they are not opposed to the granting of such an export license.

I am completely certain that you should be planting Pioneer hybrid seed corn in a rather major way in Bulgaria in 1958, and I will come to Bucharest to meet you to encourage you to immediately place your order. The history of the use of Pioneer in Rumania is pertinent to the discussion of what course you should follow in Bulgaria. In 1956 the Rumanians planted a thousand tons of Pioneer hybrid seed corn. In 1957 the Rumanians planted nearly twenty-five hundred tons of Pioneer hybrid seed corn, and they also planted a considerable quantity of Pioneer single-crossed foundation stock for the production of Pioneer hybrid seed corn within Rumania itself. They expect to harvest actually this year twelve hundred tons or fifteen hundred tons of their own production of Pioneer hybrid corn. They are supplementing that by purchasing thirteen hundred tons of Pioneer hybrid seed corn from us for planting in 1958—and they are purchasing enough single-crossed foundation stock so that they can produce four thousand tons or five thousand tons of Pioneer next year in Rumania, which they expect to supplement by purchasing more corn from us a year hence to supplement their production.

As you know, their results have been very excellent and the use of Pioneer hybrid seed corn has been one of their investments that showed not only the greatest profit, but the most rapid profit. I would guess that the average increase of the Pioneer hybrid corn over their regular varieties has been in excess of 35 percent on yield for a two year period, which is a perfectly normal expectation.

I am bringing with me a full list of the Pioneer hybrid seed corns which are available for export, and the contract made out in blank. I am also bringing a list of the single-crossed Pioneer varieties available. It has occurred to us here that you might want to get started at the production of very small amounts of Pioneer hybrid seed in 1958—a practice amount.

It has been my feeling that ultimately Eastern Europe should and will produce its own hybrid seed corn. However, I think you gain experience best and most economically by purchasing Pioneer hybrid seed corn for the original plantings you make—and perhaps get a little experience in the production of Pioneer hybrid corn yourself—and then gradually grow into your own production over a period of several years.

Both in the purchase of hybrid corn to plant for next year's crop, and in the purchase of foundation stock for the production of hybrid corn yourself, it is especially important for you to understand the great importance of purchasing the highest possible quality at a fair price, rather than ordinary or even inferior hybrids at a cheap cost per bushel.

This is a well established fact in the United States. Our own company has always produced only the finest and has always charged domestically to the farmers of the United States the highest price of the industry. We have grown from 300 bushels in 1930 to more than 500,000 bushels this year. We have grown each and every year. We have done it only by furnishing the very finest seed corn that is possible to produce. We have watched many competing companies go out of business and fail because they tried to cut the price and sell their product on the basis of a cheap price per bushel. When they cut the price, they had less money available for research—less money available for detasseling, less money available for proper drying, sizing, treating, etc.—and they finally ended up by having corn that performed less well for their customers. So, they cut the price further and their quality went lower and lower—and they finally got out of business.

The hybrid seed corn industry must absolutely be based upon the quality of the product, not the price per bushel. I am sure that cheaper American hybrids than Pioneer are available if you considered only a per-bushel price.

But, I am positive that no other hybrids are available that will bring you such a handsome profit as Pioneer. You may be interested in this respect in knowing that the total amount of Pioneer distributed in the United States is now 500,000 bushels greater than it was five years ago.

Garst & Thomas are associated in a business way with the Pioneer Hi-Bred Corn Company of Des Moines and also with Pioneer Hi-Bred Corn Company of Illinois and of Indiana—all of us produce Pioneer hybrid seed corn and the total amount produced this year will exceed two million bushels. . . .

My experience in the Soviet Union, in Rumania, and in Hungary gives me a wide background on what to expect and which corn will give you the most satisfactory results.

It will be a pleasure for me to tell you about the whole American hybrid seed corn industry development and to discuss with you other American agricultural practices. . . . It is because the American experience can be so helpful to your government that I am so extremely anxious to see diplomatic relations reactivated between our two countries.

I think the establishment of some trade between the two countries will be helpful toward the reestablishment of diplomatic relations. It will prove to your government how important it is to take a look at American agriculture. And it will prove to the United States government the advantages of reestablishing diplomatic relations at the same time. . . .

Very sincerely yours,
Roswell Garst

September 4, 1958

To Dr. Peter Voutov,
Permanent Representative of
the Peoples Republic of Bulgaria
to the United Nations,
New York 21, N.Y.

Dear Dr. Voutov,

. . . I am certain in my own mind that I can be helpful in establishing more cordial relationships between my government and your government and particularly so if I can have your cooperation. You undoubtedly know Minister [Silviu] Brucan of the Rumanian government because he is also their permanent representative at the United Nations.

When I visited Rumania in the fall of 1955, I was one of the first Americans to receive a visa from the Rumanian government. At that time, the Rumanian government had the American legation in Bucharest confined to the immediate Bucharest area. The American State Department had the Rumanian government confined to an equally small area around Washington.

The Rumanians disliked the American legation—and I think the dislike was completely mutual: the American government seemed to dislike the Rumanians. It was a most difficult situation. I scolded the Rumanians and I scolded the Americans. Because I am a farmer—and not a diplomat—the Rumanians accepted my scolding with good humor and because I am an American citizen, the American State Department accepted my scolding not with the best of humor, but with good grace.

I told both sides I felt like "an ice breaker." I thought if I broke the ice out of the river, the friendliness would increase and warm the waters—and that everyone would be happier with the thawing of the ice. And that literally proved to be true. I shall always be proud of the contribution I have made to the better understanding between Rumania and the United States, and I believe that I am in a position to make the same kind of a contribution to the relationships between Bulgaria and the United States.

For a couple of years, the United States did refuse to validate my passport for a visit to Bulgaria. But this spring they did validate David's passport so that he could visit Bulgaria—and they did agree at the same time, to issue visas for a small delegation of Bulgarians to visit the United States and see our farming operation, etc.

Two years ago, you will remember that I tried to get permission for you to come out and visit my farm and this request was refused by the State Department. Now, however, they are quite willing to have you come and visit. It certainly shows some progress on the part of our State Department and that is hopeful. . . .

We will be in full swing at harvesting corn and you, yourself, can see the most modern American farm machinery in operation, and I think it will be helpful to you and to your delegation if you have come first and really had a good look at what they will see.

I am so extremely anxious to have you come next week for a couple of days that I am going to close this letter by sending you my very finest regards and sending it off airmail and special delivery, but I am also going to wire you and I am enclosing a copy of the wire herewith. . . .

Very sincerely yours,
Roswell Garst

October 27, 1958

From Peter Voutov
Permanent Representative
of the People's Republic of
Bulgaria to the United Nations
New York

Dear Mr. Garst,
Concerning your request if we could find out whether our Ministry of Agriculture will be sending a delegation over here, it gives me great pleasure to inform you that such a delegation has been set up and it can leave for the United States any moment after it receives the American visas. Applications for visas were made on October 21, 1958 at the American Embassy in Paris.

The composition of the delegation is as follows: Liubomir Konstantinov—Deputy Minister of Agriculture, Nencho Angelov Nenchev—Head of Department "Cattle-breeding" at the Ministry of Agriculture and Kiril Pavlov Kiriakov, professor at the Higher Agricultural Institute "Vassil Kolarov".

Sincerely
Dr. P. Voutov

January 5, 1976

To Dr. Nicola Tomov
Maize Research Institute
Unega, Bulgaria

Dear Dr. Tomov:
Thanks for your New Year's greeting, which arrived today. I have pleasant memories of Bulgaria. Mrs. Garst and I were visiting with Nikita Khrushchev at Sochi on the Black Sea the day the United States and Bulgaria normalized diplomatic relations [in 1959] and he, Khrushchev, told us. So, because we had planned on visiting Rumania, we went to the Bulgarian Embassy in Bucharest and got a visa to visit Bulgaria.

It was late March, planting of tomatoes was being started, and we had a very pleasant several day visit.

We both remember our trip with pleasure. . . .

It seems to me Bulgaria has been missing a number of opportunities for more rapid progress. I feel sure that Bulgaria should not miss a single year of having several *small* delegations come to the United States to see the new things that have let us make greater progress in agriculture.

I enclose an article I wrote all of our employees and I enclose an article written for *Science*, a magazine about recent improved methods.

Also, I enclose some bulletins we have put out.

I urge you to interpret all of the enclosures and then tell the Bulgarian Minister of Agriculture that I specifically invite you to come over next spring or summer to study our most modern methods. Bring not more than three other people with you. Four people are a convenient number of any delegation. Only one of the four needs to speak English.

I do hope you can come next spring or summer.

Sincerely yours,
Roswell Garst

Yugoslavia

October 9, 1957

To Ambassador Leo Mates
Federal Peoples Republic of Yugoslavia
Washington, D.C.

Dear Mr. Ambassador,
From the experience which I have accumulated in dealing with your neighboring country of Rumania—and the experience gained by my two sons, Stephen and David, and from the experience of a local mechanic from Coon Rapids, Mr. Harold Smouse of the Grettenberg Implement Company—I am quite sure that I can make suggestions that will be of high value to your country, and as Mr. Maguire told you, I am hopeful that I will be able to visit your country in the not too distant future.

The efficiency of American agricultural production has been increasing very phenomenally in recent years. It is not only the efficiency in yields per acre of grain crops—it is the efficiency and the reduced number of man hours per unit of production of grain crops. The increased efficiency is almost beyond description.

When I went into the business of producing seed corn twenty-seven years ago, practically all of the corn planted in the United States was picked by hand. It took an average approximately eight minutes per bushel to just pick the corn. Now the average for the state of Illinois is a total expenditure of six man minutes to plow the ground, disk, harrow, plant, rotary hoe, cultivate, spray, pick, and crib a bushel of corn. Actually under the most favored circumstances, we have been able to raise and pick and crib corn for less than 3 minutes of man labor per bushel of corn. . . .

A few years ago—say ten years ago—it took 4½ pounds of feed to make one pound of broiler meat. Five years ago it took 3½ pounds of feed to make one pound of broiler meat. Now it takes on the average, about 2.7 pounds of feed to make a pound of broiler meat. A similar reduction percentage-wise has been made in the amount of feed required to produce a dozen eggs.

An almost identical percentage increase in the efficiency of the feeding of swine and chickens has been the discovery and widespread use of not only antibiotics but an exact protein balance, an exact mineral balance, and the additions of vitamins. Actually, egg production in the United States is just as high if not higher in the winter months than in the summer months.

In two years of traveling in Eastern Europe, I have come to realize the reasons why the United States agriculture has progressed at so much more rapid a rate than the agriculture of Eastern Europe. I am not in any way critical of your present position—nor boastful about our own very great efficiency—I know the reasons. . . .

Particularly in the Danube Valley countries such as Yugoslavia, Rumania, Hungary, and Bulgaria, corn is historically one of your major crops, and so your first place to start, of course, is with corn. That calls for superior hybrid corns and mechanization and fertilization and irrigation and insecticides and herbicides.

I have known for some time that you were using hybrid seed corn in Yugoslavia. I think the results obtained from the use of hybrid corn in Yugoslavia are not as good as the results obtained by Rumania, and I think the reason is very simple. The hybrid seed furnished to Yugoslavia was, in most cases, not first class hybrid seed—in most cases, it was indeed rather poor hybrid seed—which was too bad. I wish to tell you how this came about.

Under the American assistance program, the funds appropriated for assistance to Yugoslavia were made available by our Congress, and in agricultural matters, the United States Department of Agriculture assisted in the purchase of farm supplies. This seems like an intelligent way to handle the matter, and I am not critical of your government for believing that it was an intelligent way to handle the matter.

But the United States Department of Agriculture, when they set out to purchase seed under the assistance program for use in not only Yugoslavia but in Greece and Italy and France and West Germany, simply sent word out to the hybrid seed corn industry that they wanted to purchase 500,000 bushels of hybrid corn of several different maturities, and that while they wanted different maturities such as early, medium, and late, they were not very careful about any other requirements. The germination only had to be 85 percent, and a 15 percent tolerance was allowed, so the germination could be as low as 75 percent and still have it qualify. It became a business of simply selling the lowest quality product of the American hybrid seed to the U.S. government for European assistance on a price basis—and not on a quality basis. I protested with the Department of

Agriculture at the time that they should buy hybrid seed corn on a quality basis, but they were not equipped with personnel that would let them do so and proceeded to do as they had been doing. . . .

I understand that in Yugoslavia you have had difficulty in getting the peasants to plant hybrid corn on their own farms because they had not had satisfactory experience with the hybrids. In Rumania, exactly the opposite thing is true—our Pioneer varieties did such grand jobs in Rumania that the Rumanian government actually had difficulty in keeping the peasants from stealing the hybrid corn that was being grown for planting the second year because it was so attractive and so beautiful.

Last spring, in fact, the Rumanian government did purchase Pioneer hybrid corn from us and break it up into smaller sized bags and sell it to the peasants for planting—and our present negotiations with the Rumanians include the thought that a fair percentage of the Pioneer they purchase this winter will be sacked in small sacks especially for distribution to the peasants.

The difference in results that can be obtained from the highest quality of hybrid seed corn and the results obtained from ordinary or inferior hybrids is very great. Farmers in the United States would no more think of buying a 1950 model Pioneer hybrid than they would think of buying a 1950 model automobile. Constant research brings constant improvement.

I am extremely anxious that your government buy five hundred to one thousand tons of Pioneer hybrid seed corn from our company for comparison purposes for 1958 planting. It will, I am sure, bring you far superior results to any hybrids which are being produced or have been sent to Yugoslavia in the past and will give you a new vision of what is possible.

We will offer to sell you Pioneer hybrids at the same price we have been selling them to the Soviet Union and to Rumania and Hungary—that is, five dollars per bushel for the medium and late maturing varieties and six dollars per bushel for the early and extra-early varieties. Because of our experience in Rumania, we are sure of the adaptation. You can use medium and late maturing varieties in the low altitudes along the Danube River— you will need the early varieties in the higher altitudes and perhaps some extra early for your highest altitudes.

So much for the hybrid corn.

It seems to me that you should probably also buy a few complete sets of the most modern American farm machinery for at least demonstration purposes, and I will be happy to discuss this whole mechanization angle with you.

And finally, you certainly should buy at least a limited amount of the very finest of American fertilizers, again as a demonstration. And you should buy some American insecticides and some American herbicides, again for demonstration purposes. And if you do not have some irrigation started, you should within the next year get started on some irrigation.

To demonstrate how to raise a maximum corn crop with minimum effort of human beings is really a very simple undertaking. If you have some large state-owned farms—which I presume you do—or large collectivist farms, those are the places to put on a demonstration. In a country as populous as Yugoslavia and as widespread as Yugoslavia, you probably should have several demonstrations in various sections of your country. There is, however, no difficulty whatever in putting on a demonstration with assured results. The methods are not complicated—all of the facilities are completely tried and proven—and it's simple. . . .

I do not wish to encourage Yugoslavia to spend more money for improvements in agriculture than your government feels is justified in being spent, other matters being taken into consideration. But I do find it necessary to point out that it would seem to me that anything up to 1 dollar per person would not be an excessive amount to allocate to such a very pressing problem. And I do wish to point out that probably an expenditure of a minimum of twenty cents per person or thirty cents per person in the next year might prove to be the most beneficial place that you could use funds.

It is for the above reasons that I am so happy that you [are coming] to Coon Rapids. The stories that I could tell you about the extreme efficiency of American agriculture will sound very greatly exaggerated anywhere else. They will be in no way exaggerated, as I can clearly show you. And I want to discuss with you while you are here the full gamut of opportunities that present themselves to your country to increase its agricultural production. . . .

Very sincerely yours,
Roswell Garst

December 20, 1957

To Mr. Leo Mates
Ambassador from Yugoslavia
Yugoslavian Embassy,
Washington, D.C.

Dear Ambassador Mates:

I have just returned from a short trip to Eastern Europe. I stopped a couple of days each in Hungary, Rumania, Yugoslavia, and Czechoslovakia.

In Yugoslavia I visited with Mr. Ivan Bukovic, Mr. B. Zlataric,[1] a couple of your best corn men . . . and others, at great length about the seed corn and I visited with Mr. Slavko Komar, your minister of agriculture at even greater length about the more efficient use of corn in the feeding of chickens, pigs, and cattle.

Your country has produced this year some sixteen thousand tons of hybrid corns that were developed by the several American colleges fifteen years ago. These hybrid varieties are bringing satisfaction in your country, compared with the open pollinated varieties—which of course is to be expected. But they are obsolete by present American standards. It would be like manufacturing 1940 automobiles: 1940 automobiles are far more satisfactory than horses, but not nearly so satisfactory as the 1958 automobiles.

It looked to all of us as though the best thing to do was to get some good comparisons between the newest and finest Pioneer hybrid varieties and the varieties currently being raised by your government, and for this purpose I recommended the purchase of five hundred tons or perhaps even as much as one thousand tons of our best Pioneer varieties. I will be corresponding directly with your government just as soon as I can compile the comparisons that I have agreed to furnish them, which probably will not be until the first of next week. I shall be sure to send you a copy of the letters I write to them.

But, at the moment, I did want to inform you that I had returned after a highly satisfactory trip.

And I did want to thank you for having come out to visit us here at Coon Rapids and for having reported to your government about your visit.

And, I did want to especially extend to you and Mrs. Mates the

---

1. Agricultural officials.

warmest season's greetings and best wishes, not only from me, from Mrs. Garst, from Mr. and Mrs. Stephen Garst, and from Mr. and Mrs. David Garst, from all of us here at Coon Rapids, who so much enjoyed your visit.

Very sincerely yours,
Roswell Garst

January 20, 1958

From Slavko Komar
Member of the Federal Executive Council
Secretary for Agriculture and Forestry
Belgrade

Dear Mr. Garst,

I want to thank you first of all for the very interesting material you have sent me, "The World Shortage of Protein Feed for Livestock and Immediate Prospects for Improvement."

Both from your material and your very interesting statements you made while you were here I am getting more and more convinced that in your country a great progress has been made in production and utilisation of corn as a crop.

I would be very obliged to you, dear Mr. Garst, if you could send me some additional material and give me further information of your experiences on this matter.

I am taking this opportunity to ask you to meet Mr. Vladimir Trifunovic, specialist of the Institute for Plant Breeding and Plant Production, Zemun (near Belgrade). Mr. Trifunovic is going to stay at the Agriculture College, Urbana, Illinois to the end of this year. During his stay at the College he will write to you for the arrangement of his visit to you and your very respectable company.

Very sincerely yours,
Slavko Komar

January 31, 1958

From Ivan Bukovic
President, Institute For Plant Breeding
& Plant Production
Belgrade

Forgive me for not answering sooner, as it was necessary to consult with our production organizations and experts, being that your offer came after we during your stay here declined it, and we had to take up the matter again for reexamination.

You will understand that it is hard for us to enter into purchasing of hybrid corn considering the high yield per hectare successfully accomplished here. This we emphasize not because we believe that we do not have to do better, but because hybrid which you are offering places us in a position that we buy seeds from you every year and you do not give us a chance to master our own process of production, which for us would be interesting only in as much as your corn would give exceptionally bigger results.

Because of this we decided for now to import only 3 tons of your hybrid seed, with intention of study and next year we could probably enter into further business arrangements.

Once more I thank you for your efforts and kindness that you have shown on your visit to Yugoslavia, discussing with us problems of progress in corn production and its use for broader purposes. I wish to remind you of our agreement which took place of discussion in two directions:

1. We are willing to send a group of two people [to] your farm to stay there few weeks to get acquainted with process of work and production on your farm. We ask you kindly in this regard to give us conditions for these people to stay on your farm.

2. I have asked you to recommend for me and my organization an adviser for modern production of cattle, hogs and fowls with consideration that the expert stays with us in Yugoslavia one year and to be compensated by us. His work would consist of advice and assistance to our large estate producers in overcoming [sic] modern methods of production. Especially I have in mind feeding of cattle by way of UREA [sic] that we have discussed.

I hope that you are well, best regards from me and my associates with whom you got acquainted while visiting here.

Ivan Bukovic

February 22, 1958

To Ambassador Leo Mates of Yugoslavia

Mr. Philip Maguire

Gentlemen:

I am enclosing a copy of a long cable I have just sent out to Mr. Ivan Bukovic that is self-explanatory.

I haven't been able to sleep well since I received their order for three tons of Pioneer corn. We are sending them eight different varieties, and three tons only amount to 120 bushels—so we are sending only 15 bushels of each variety.

For a great corn growing country like Yugoslavia to be experimenting with such a very small amount of such very superior hybrids is nothing less than a tragedy! They can do nothing but put it in a few experimental plots. If they get the finest kind of results in experimental plots, they still won't know anything much—they won't be in a position to order substantial quantities. It takes field tests—and widespread field tests—so that you get to see the whole quality of the corn—its ability to stand straight, its ability to mature, all of the factors involved—before you secure real knowledge.

While we have agreed and will of course deliver the three tons they ordered, I feel absolutely certain that it will be a wasted effort on their part and ours. If they only get the three tons of corn I would feel almost certain that their information will be so sketchy and so thin that whatever they order next year they will order without any confidence, and I would almost guarantee that if they only get three tons this year, they will need to look again and that they won't buy more than fifty tons next year—and they will do that with hesitancy—and that they will lose several years in continuing to produce obsolete models of hybrid corn.

If they buy the three hundred tons, which amounts to less than 2 percent of what they did actually produce this year, they will get a wide-spread test of two varieties that are extremely well proved and extremely popular in the most intelligent corn growing area in the world—that is, extreme northern Iowa and southern Minnesota, Wisconsin, extreme northern Illinnois, Michigan—the northern Corn Belt.

Ambassador Mates, I think you must realize that with an institution of the size of Garst & Thomas I could not afford to make the cable I sent and this letter as strong as I have made them if I were not completely sure of my judgment in the matter. No man in the United States has had more experience in the hybrid seed corn business than I—and not only in the United States but no one from the United States has had wider experience in Eastern Europe than I.

And no one has enjoyed visiting in your area more than I—and no one is more anxious to be helpful to your country than I.

But very frankly, my mature judgment is that I can spend far more of my time helping the Rumanians, helping the Hungarians, helping the Soviet Union, helping the Czechs than I can spend it helping the Yugoslavians if the Yugoslavians are going to be as hesitant and as afraid as their actions up to now have indicated.

I will make it very frank!

I have always considered that when I sold a customer in the United States seed corn I was entitled to help him in every way, learn how to raise it with the greatest mechanical advantage and how to feed it with greatest efficiency. I have thought—I still think—and always will consider that this is part of the service that the customer should reasonably expect from me— part of the service that I ought to give him. No day has been too long, and I have never tired of giving my customers in the United States the very finest possible advice. I have considered it a part of my duty.

And I have taken exactly this same attitude toward foreign customers. And I expect to continue to take the same attitude toward foreign customers.

In the United States, I have done what I believe any businessman does—I have not refused to give service to noncustomers and especially if I thought good service in the way of advice would encourage them to become customers. But I have always spent a lot more time with good customers than with noncustomers because I thought it was my duty to do so.

If the Yugoslavian government only buys three tons of Pioneer for planting in 1958, I see no likelihood they will buy substantial quantities in 1959 or any other year, and I am very frankly not going to worry half as much about being extraordinarily helpful as I will if they buy three hundred tons because if they buy three hundred tons this year, they are quite likely to buy one thousand tons next year—and it puts them in a class of being my most favored customers.

That doesn't mean that I won't be happy enough to have a couple of their citizens over here for a few days even if they only buy the three tons. But it does mean I am not going to "roll out the red carpet" if that is a good expression to use—and really exert myself to the extreme as I will with big potential customers.

My whole family, Ambassador Mates, enjoyed you wonderfully when you were out here. We hope you come back any time you wish to see American agriculture or find out about it. We will, of course, not refuse to be helpful to any people coming over from Yugoslavia. But I just wanted to emphasize that it is only natural for business people to furnish the greatest services to the best customers.

I think the order for three tons was ridiculously small!

When I was asked to come to Yugoslavia to visit personally with Mr. Bukovic, I refused to go until and unless you had come out here because I wanted to be sure that your government was informed by you that Garst & Thomas were really a substantial outfit—that the Garst family really knew about agriculture other than seed corn—and that we were friendly, cooperative people. I feel sure you must have reported these things to your government.

And with that background, I did go—I did take time out of a very busy winter—and I did give the best advice possible, and I did offer to continue to give advice and to continue to be helpful.

When your government ordered three tons of Pioneer hybrid seed corn, it was like a "vote of no confidence." I could interpret it no other way. That is why I have not slept well. I am not used to votes of "no confidence." If I were prime minister and got that kind of a vote, I would of course, resign.

That's why I sent the cable which is enclosed. That's why I have written you at this great length.

If you feel like doing so—and only if you feel like doing so—I wish you would either telephone your government or cable them your best advice in this matter. I know your country is not long on dollars. That's why I suggested as little as five hundred tons or eight hundred tons in my letter to them. That's why I am now willing to suggest as little as three hundred tons. In a broad way, the three hundred tons would only cost $60,000—a small fraction of one cent per person living in Yugoslavia—an absolutely insignificant amount for a corn growing nation to spend to find out how to produce 10 percent more corn than they are now producing with the same effort and with the same expense. That's why I feel so strongly.

Three hundred tons even is such a small amount that I consider it to be about the minimum that will prove my point.

The letter gets too long—but I can sleep better after having written it. Not to have written it would make me feel I had not done my part, feeling as I did. I don't want to feel that way. With kindest regards to you and Mrs. Mates, I am,

Sincerely yours,
Roswell Garst

March 11, 1958

To Ambassador Leo Mates,
Yugoslavian Embassy,
Washington, D.C.

Dear Ambassador Mates,

. . . There is one thing I cannot understand about all of the countries of Eastern Europe. That is the lack of correspondence. I think it's a serious mistake and I can't help but think that it is a policy decision which should be reversed. And it seems to be universally a policy decision.

I have never had a letter—not one single letter—from any delegation that has ever visited here, from Russia, Rumania, Hungary; not an inquiry as to what new improvements we may have discovered; not a question about the problems they face. In fact, the only letter I have ever received from any of the area was a letter of appreciation I received from Minister Komar thanking me for some material I had sent him.[1]

While I am very anxious to be helpful, I don't know how to be helpful to people who do not ask questions. I can't help but think that the greatest kind of thirst for knowledge about the most modern American methods in your country exists.

President Bukovic took me out one Sunday morning to a meeting at a cooperative farm, where I met with some twenty agricultural leaders, and I spent some four of five hours answering questions about every type of operation about Corn Belt agriculture. I know I could not have satisfied their minds. I know they must have many things that they would still like to know. I'd be most happy to answer questions by mail. But I don't get the questions by mail.

I think Yugoslavia is missing much by not encouraging more correspondence. . . .

Sincerely yours,
Roswell Garst

---

1. Garst was overstating the case. He did receive other letters.

March 18, 1958
From Leo Mates
Ambassador of Yugoslavia
Washington, D.C.

Dear Mr. Garst:

I have seen your letter of March 11, 1958, only today, due to a prolonged absence from Washington.

I fully understand your misgivings about the lack of correspondence from Belgrade. I wish to assure you, however, that this is unfortunately no policy decision, but a regrettably bad habit of our people, which I have experienced myself quite often. I said "unfortunately" because if it were a policy decision, it could be just the reversed [sic], but to cure this bad habit is frequently much more difficult. I hope your letter might influence those concerned to attempt to cure this bad habit.

I was very glad to learn that you contemplate a trip to Yugoslavia, you or somebody from your organization, and I think that it would be excellent if Mr. David Garst could come.

Having said what I said about bad habits, I wish to add that we have some good habits, too, and that is to appreciate good contacts and oral exchange, so that it would be wrong to deduct [sic] from the lack of correspondence a lack of interest or to anticipate anthing but friendliness and cordiallity [sic] towards everybody who comes to Yugoslavia. . . .

Yours truly,
Leo Mates

April 4, 1958
From Ivan Bukovic
President, Institute for Plant Breeding
and Plant Production
Belgrade

Dear Mr. Garst,

I got your letter for which I am indeed very thankful to you. I am sure you would understand that we could not buy a greater quantity of corn seed than we did e.g. 3 tons. We shall put the seed into exact tests and make comparisons with the seed we already have in our country. On the basis of obtained results we will economically appraise whether it should be rational to buy from you the seed material, or stay with the old one—or try to find other ways to improve the corn seed. You must understand us because business is in question and you are a business man. My only wish is, that the established business cooperations between us should be of a mutual interest. If either of us should suffer from losses it might disturb our friendship and business relations we established.

I read with great care the various suggestions you lay out. I entirely agree with you that a group of Yugoslav corn seed producers and at the same time a group of experts for meat production visit you. Meanwhile, we must consider yet whom to send and at what time to come to the USA. It is of great importance to select the best people because we want to have the maximum benefits from their dwelling in the USA.

You must understand, that we here in Yugoslavia are interested in the revolutionary process in the production of livestock particularly in production of meat starting from chickens to hogs and cattle. I think that in this question the farmer of USA reaches more than any other and that your experiences are the most precious one. Therefore, I would appreciate very much if you would send us the names of the best known specialists and scientists who deal with the above mentioned problems in the USA, in order to get from some of them scientific works and various periodicals and bulletins, and to get others through the Food and Agricultural Organization (FAO) who would be willing to come to Yugoslavia to organize for us some dozens of modern farms for the production of meat and eggs.

We are also interested in the modern solution of dairies, as well as in specialists who gained high results. Therefore I beg you, if it does not represent any trouble for you, to turn our attention to the best technological solutions in the USA, as to the existing ones which are on practice already, as well as for that which are obtained and not yet put into practice. Further on, would you kindly help us to get some bulletins and periodicals through which we could learn such things. We would appreciate very much to subscribe [to] them continually. At the same time we wish to get names of persons who would like to come to Yugoslavia and help us in that great job. . . .

Once more I am very thankful to you and wish our cooperation should continue to a mutual benefit, as well as in the interest of a better and more cordial acquaintance and cooperation between our and your people.

Sincerely yours,

Ivan Bukovic

April 28, 1959

To Mr. Slavko Komar,
Minister of Agriculture,
Mr. Ivan Bukovic, President
Institute for Plant Breeding
and Plant Production
Belgrade, Yugoslavia

Dear Mr. Komar and Mr. Bukovic,

I was terribly sorry to have missed you both when I was in Yugoslavia, but I had Mrs. Garst along with me, and she had been on a very long trip throughout the Mediterranean area and throughout the countries of Eastern Europe, and by the time we got to Belgrade, we were tired and in a hurry to get home.

There are many things I want to write you about and many reasons why I wanted to see you. Probably first and foremost was that I know you are building a plant for the production of urea in Yugoslavia. I know that urea can be one of the greatest instruments of agricultural production that any nation can have. Urea can not only be used for fertilizer but it can also be used for the protein of ruminants.

But believe me, you need to really know how to use urea to make it effective, and the easiest way to learn to use it is by the use of portable feed grinding mills such as my son, David, talked to you both about when he was there in June of last year.

Furthermore I am not at all sure that you should not have a couple of sets of really modern corn growing machinery for use in Yugoslavia.

I know you had been purchasing some machinery in England and other places, but when it comes to big corn growing machinery that will let six men completely cultivate eight hundred hectares, I do not believe that you have seen any of it.

The Soviet Union is ordering four such sets for delivery this summer, the Bulgarians have expressed an interest in ordering two such sets for delivery this summer, and the Hungarians, I believe, will buy at least one set for use in Hungary.

In as much as Mr. Grettenberg, who has suppled these countries with their machinery before, has just been making up lists of such sets of machinery, I asked him to list two sets with the idea that this list would be forwarded to you by him for your consideration.

You will note that the last item on the list is the two portable feed grinders.

If I were going to give you advice, it would be to order two complete sets of farm machinery as listed above, and then I would personally urge you to buy fifteen or twenty portable grinders as listed. They are equipped, as David told you, with the molasses applicator, and all you will need to do is mix the urea with the molasses and applicate corn cobs or corn stalks or most any cheap cellulose and feed it to your cattle, and you can winter cattle cheaper and easier and better than you have ever wintered them before.

The reason I am writing you and urging you to place an order for what machinery you want at the earliest possible moment is that the demand for farm machinery in the United States is so high that the International Harvester Company will not agree to supply the machinery less than sixty days after the date of order, and they do not agree to complete the deliveries for ninety days after the date they are ordered. The same general type of thing exists with the Melos grinder people—that is, they have a big demand in the United States and will want sixty to ninety days in order to get the machines fabricated.

When you add a month's time for transportation on top of that, you get the earliest possible delivery as being toward the [first] part of September.

I think any machinery you order should have at the same time an order for 10 percent of the value of the machinery in repair parts, and you must give this consideration as you place the order.

If you are interested in ordering the suggested list of machinery and particularly if you are interested in ordering the extra grinders that I have suggested, if you will let me know just as soon as possible, I will have Mr. Grettenberg prepare a contract covering your wishes. . . .

Sincerely yours,
Roswell Garst

Czechoslovakia

May 21, 1957
To Dr. Karel Petrzelka,
Ambassador from Czechoslovakia,
Washington, D.C.

Dear Ambassador Petrzelka,

First I wish to tell you how much we enjoyed having you and Mr. [Roman] Skokan visit with us last week. I am sorry that rainy weather prevented field work so we did not get to show you the actual operation. But we did get to have you visit with practical farmers like Steve and Dave and John Chrystal, and we did get to show you the method that is letting American agriculture produce excess food of the highest quality with the lowest possible man hours.

And I did get to discuss with you at length the opportunities that face Czechoslovakia—not only the opportunities to increase your own food production per acre but to manufacture the tools with which neighboring countries can increase their food production per acre. Yours is a largely industrial country—one of the finest in all the world—and there is going to be a terrific market for all sorts of modern agricultural equipment on a worldwide basis. . . .

So Czeckoslovakia needs to learn thoroughly and immediately not how to raise corn alone—but how to transfer that corn into human fats and human protein and hides—and you need to learn the lesson so well that your country can teach other countries, indeed, to help furnish the equipment for the other countries as they learn the lesson.

Aside from the United States there is no area where the production of food is anything like completely mechanized. Sure they combine wheat in Canada and Australia—but in a broad way, most of the world's food is produced without mechanization or at least not nearly as complete mechanization as we have—without the coordination of all of the branches of the Industry.

. . . The more I think it over, the more I believe that Czechoslovakia should immediately lay plans to go the whole distance so far as projects are concerned. Certainly your largest single expenditure for the 1958 season should be for enough of the best American hybrid corn to plant a large part of your nation.

The profit from hybrid corn is perhaps the biggest single profit to be obtained. A 35 percent increase in yield is assured as a minimum. Actually, the increases were much higher than that last year in the USSR, Rumania, and Czechoslovakia.[1] The profit is just so large your country cannot afford not to go the whole distance on hybrid corn.

And you can use hybrid grain sorghums very effectively and with equally great profit in southeastern Czechoslovakia—the Danube River valley just north of Hungary. . . .

I hope that you will report to your government about the trip you and Mr. Skokan made to Coon Rapids and tell your government that I shall be happy to arrange for a visit by one or two or three of your top agricultural people from Czechoslovakia.

Actually, some responsible man from your Ministry of Agriculture who has broad vision—someone whom all of your ministries have confidence in ought first to come over for a look.

Technicians can come later to study individual projects. The first man that comes over should be a man of authority and respect in your country to take a broad look. He does not need to stay a long while—two or three weeks is enough—and I shall be most happy to do the best possible job of being helpful. . . .

> Very sincerely yours,
> Roswell Garst

November 14, 1957

To Dr. Karel Petrzelka,
Ambassador from Czechoslovakia,
and Dr. Roman Skokan,
Commercial Attaché,
Czechoslovakian Embassy,
Washington, D.C.

Dear Dr. Petrzelka and Dr. Skokan,

. . . When I was in Prague last winter, I did discuss with your minister of agriculture the advisability of planting rather substantial amounts of

1. Undoubtedly Garst means Hungary.

Pioneer hybrid seed corn in Czechoslovakia. And I did discuss with you two gentlemen when you were here, the advisability of not only increasing your corn production but also the advisability of getting started at the chicken business.

And Mr. Finley and Mr. Tereshtenko did call on your minister of agriculture and discuss chicken equipment and feed mill equipment, and I know that he has invited several of your agricultural people over to study the matter of chickens and feed equipment, etc.

In the case of Pioneer hybrid seed corn however, I am somewhat distressed by the delay for which I am probably responsible in getting an order covering the needs of your country for 1958 planting.

In Eastern Czechoslovakia, corns of early maturity are very suitable. For instance, in the 1956 tests—tests which I saw a year ago—both very near to Budapest and also the tests in the Lake Balaton [region] of Hungary, Pioneer 352, Pioneer 347, Pioneer 344, and Pioneer 349 all showed very excellent results. These varieties are still available in decent supply and do not cause me the immediate worry that the extra early varieties cause me, and it is the extra early Pioneer varieties that are most suitable for the higher altitudes of Western Czechoslovakia.

As I perhaps explained to you, only a very small percentage of the total corn grown in the United States is of extra early maturity. But a great deal of corn in Europe is of extra early maturity. For instance, all [of] Poland needs extra early maturing varieties. The same thing is true of Holland—and Denmark. And the same thing is true with the high altitudes even in Rumania—when you get up in the Carpathians.

Because all four of Mrs. Garst's grandparents came from Bohemia, I have a little more sentiment connected with Czechoslovakia than with any other nations of eastern Europe—and I am ever so anxious to be helpful to your country in any way possible . . . .

It is my intention to go to Europe in early December and to call on your minister of agriculture while I am there. I must first go to Rumania, and I expect to talk to the Rumanians, Bulgarians, Hungarians, and Yugoslavs before I reach Czechoslovakia. I feel absolutely certain that unless your order is placed for the extra early varieties they will not be available by the time I get to Czechoslovakia. . . .

Anyhow, I will discuss this whole matter with either one or both of you when I hand you this letter. I will also discuss the supply of Pioneer hybrids available to you and your country—and urge that your minister immediately place an order for at least approximately the quantities he wishes—and I feel that this is especially necessary in the case of all of the extremely early varieties which are so necessary for higher altitudes of Western Czechoslovakia.

Very sincerely yours,
Roswell Garst

December 20, 1957

To Ambassador Dr. Karel Petrzelka
Czechoslavakian Embassy
Washington, D.C.

Dear Ambassador Petrzelka:

I just wanted you to know how much I appreciated your kindness in arranging for me to pick up my Czechoslovakian visa in Bucharest on my recent visit to your country. The matter had already been arranged and I had no difficulty at all in securing the visa, and I had a most pleasant visit in Prague discussing not only the increased use of Pioneer hybrid seed corn, but also in discussing the more efficient feeding of corn to swine, chickens, chickens for eggs and meat, and cattle.

Your nation is primarily an industrial nation rather than an agricultural one. So of course your nation will never be a large user of Pioneer hybrid seed corn.

But you do have some very excellent corn areas in the eastern part of Czechoslovakia, north of Hungary, and you do produce limited amounts of corn widely over the whole nation and I found your government officials most interested in maximum production compatible with the opportunity in all of the areas.

And I found them very anxious to discuss new methods of agriculture throughout the whole gamut of agricultural problems—that is, fertilizer, hybridization, mechanization, better methods of feeding—interested in the whole new and ever-improving agriculture.

I simply wanted to write you that, after many delays, because of other pressing problems which I faced, I did finally make the trip and I highly appreciate your assistance in arranging the Czechoslovakian part of it.

I want to extend the season's greetings from all of us here at Coon Rapids to both you and Mr. Skokan and urge either or both of you to visit with us again whenever you can find time to do so.

Sincerely yours,
Roswell Garst

March 20, 1964

To Ambassador Dr. Karel Duda,
Czechoslovakian Embassy,
Washington, D.C.

Dear Ambassador Duda,
We were delighted to have you and Madame Duda and Counselor Vlastimil Tuma as guests in our home yesterday—and I was happy to be able to show you a part of our farming operation. . . .
Your country is recognized as being highly skilled mechanically. But you have not had the necessity of making farm machinery that allowed you to raise maximum agricultural production with the minimum of labor which faced the United States.
We build simpler machinery—machinery that one man can operate—machinery that lets us produce on the Garst Farm one bushel of shelled corn with two minutes of man time—and I am talking about the whole season's work.
In 250 minutes we can shred the corn stalks, disc the acre, plow the acre, disc it again, harrow, plant, cultivate, spray with insecticide, harvest and store the corn from an acre—which will yield 125 bushels per acre.
The average for the whole state of Iowa is to raise eighty bushels per acre with less than three hundred minutes per acre—less than four minutes per bushel.
Other grain crops are raised in the United States with a proportionately small labor requirement. Without any doubt, the United States leads the world in the production of food per hour of man time required.
Some other countries do as well or better in yields per acre—Holland, as an example—but no country even approaches us in return per man hour.
Certainly the whole world must learn to mechanize agriculture to a much greater extent than ever before in history—and must learn at once.
In nonindustrial countries such as Central and South America and Southeast Asia and Africa, it takes up to fifteen hours of man time to produce a bushel of corn—nine hundred minutes in Central America as compared with two to four minutes in Iowa. The results—
We use 7 percent of our population in agriculture and produce great surpluses after we enjoy a very fine diet! In Central America it takes 85 percent of the people in agriculture—and they furnish a diet far too low in protein.
The rest of the world is in between . . . .
Where we have been using 7 percent of our people in agriculture, Eastern Europe has been using anywhere from 40 percent to 60 percent of

its people in agriculture. It should be possible to cut that down to under 20 percent in the next five years.

But an industrial nation like Czechoslovakia should not only be interested in reducing its own agricultural labor requirement—it should be in a position to help the less fortunate nations equip their farmers with better machinery.

The same thing that is true with farm machinery is true with American road building machinery.

The United States has more good farm-to-market roads than the rest of the world. *And no agriculture can prosper without good roads* . . . .

Again I can say that a country with industrial know-how such as Czechoslovakia should be buying the latest American farm machinery and road building equipment if for no other purpose than to stimulate your own engineers and designers into building machines which are going to be required *all over the world.*

The United Nations population studies indicate that whereas the population gain in the world was 1.4 billion in the first sixty years of this century—the population growth will be 3.4 billion in the last forty years of the century.

Every industrial nation must prepare to make maximum contributions toward feeding that population explosion.

That is why I hope you will be able to get not a large delegation from Czechoslovakia, but a small and extremely influential group to come over for a visit in late August or early September. I think such a delegation should include a man from the Department of Foreign Commerce, a man from your heavy equipment manufacturing department, and not more than one man from your Department of Agriculture.

I think they should have the authority to purchase several sets of the most modern American farm machinery and a couple of sets of the most modern American machinery for the building of "farm to market" roads.

We, here at Coon Rapids, will be willing to be helpful. . . .

Sincerely,
Roswell Garst

August 10, 1964

To Dr. Karel Duda,
Ambassador from Czechoslovakia,
2349 Massachusetts Ave. N.W.,
Washington, D.C. 20008

Dear Ambassador Duda,

It is fine to know that a delegation of agricultural machinery men headed by Mr. [Josef] Nagr—the first deputy minister of agriculture—is coming to Coon Rapids. It has surprised and disappointed me that your government has waited nine years before sending such a delegation. I did in fact, write your Department of Agriculture to send them over in both the fall of 1955—and the fall of 1956.

However, I think it is a thing you should do each year—or at least every other year. The rapidity of change is so rapid that in a five year period, agricultural machines change completely.

It seems to me that your government in Prague should authorize this delegation to *purchase* farm machines—not just look at them . . . .

So I would like to have you do two things.

First—tell the delegation to arrive by September 2. We will be putting up silage, we will be picking corn, we will be doing all the things that they should see. We will be feeding cattle on corn cobs with molasses and urea—as *your* country should, itself, be doing.

Second, tell your government that the delegation should be authorized to purchase. If they can place orders while they are here, the machinery can reach your country in time for use next spring. If they don't place the orders while here, they won't get it there in time for next year . . . .

Very sincerely yours,
Roswell Garst

Poland

September 9, 1957

To Dr. Jan M. Kielanowski, leader,
and other members of the
Polish Agricultural Delegation of 1957

Gentlemen:

I do not know whether you are authorized to make actual purchases in the United States or not. I do know, of course, that you will either be authorized to make actual purchases, or at least that you will make recommendations about purchases. So, in any case, it will be helpful to make some suggestions to you.

While I have never been in Poland, I have two different falls[1] traveled in the Soviet Union, in Rumania, and in Hungary, and I have flown across Poland a couple of falls—and I am relatively familiar with your agricultural problem.

You need to raise more total grain and particularly you need to raise more total feedgrains so that you can feed animals and poultry to give your people more protein and more fats. The people in Poland do have a diet with a sufficient number of calories. You raise potatoes well, you raise cabbage well, you probably have at least almost enough bread grains, such as wheat and rye. You can materially increase the production of all of the above mentioned grains with a more widespread and heavier use of fertilizers.

When it comes to the feed grains, such as oats and barley, and especially corn—you simply haven't had sufficient amounts available to expand your chicken production, your swine production, nor your cattle production. Swine not only produces the lean meat your country needs for protein, but it also produces fats in the form of lard, which is in such short supply over much of the world, including Poland.

I am positive that Poland can profitably expand the corn acreage, and corn is by all odds the major feed crop of the world. It can be used well not

---

1. That is the autumn of 1955 and of 1956.

only for hogs, but for chickens as well—and the cornstalks can be fed to cattle along with the corn in the form of ensilage very profitably.

Corn—even the very early varieties—will generally yield well over twice as many feed units as will oats. In the United States corn easily produces more than three times as many feed units on a given area as oats. So I am certain that all the parts of Poland where corn can be grown should see the corn acreage expanding. And when you plant corn the most profitable single investment is the use of hybrid seed corn. You need to use extremely early Pioneer hybrid varieties—and these can be supplied to you by the Pioneer Hi-Bred Corn Company of Des Moines much better than they can possibly be supplied by Garst & Thomas. . . . .

Earlier this summer Ambassador [Romuald] Spasowski and Mr. Iwaszkiewicz visited Coon Rapids, and I went over this whole matter with them, and they subsequently encouraged Mr. George Finley of the Finco Company to stop in Poland on his way to the Soviet Union. Mr. Finley has already visited Poland and is presently in Rumania, but the vice president of his company, Mr. [William P.] Ludwig, is coming out to Coon Rapids while you are here to visit with you about feed plants and chicken equipment.

The Finco Company has furnished seed corn drying equipment to both the Soviet Union and Rumania, and both countries and also Czechoslovakia have expressed interest in feed mixing plants and chicken equipment, and it now appears likely that the Finco Company will be in the best position of any American company to serve you in these matters.

My two sons, Stephen and David, whom you will meet here at Coon Rapids, each spent sixty days in Rumania a year ago with their wives. They are [in] thorough agreement that lack of efficiency in the raising of poultry and swine and cattle, as compared with American methods, is one of the real tragedies of Eastern Europe.

Of course I also expect to show you modern American agricultural machinery—how few man-hours we use and how much machinery we use in our operation as compared with your own. You have come at a very fine time because you can see some corn picking, you can see silos being filled, and you can see the terrific advantages of mechanization. . . .

A complete set of American farm machinery large enough to farm two hundred hectares costs in the neighborhood of twenty five thousand dollars and would reduce your labor requirement something terrific. In Rumania, for instance, one man fully mechanized was able to do about as much as twenty men were able to do without complete mechanization.

You will be seeing at the Garst & Thomas hybrid corn plant, the world's largest individual seed corn processing plant. We expect to produce this year more than 500,000 bushels of Pioneer hybrid seed corn. At the same time we will be producing something like 125,000 bushels of Garst & Thomas hybrid grain sorghums. It will be a matter of high interest to you, I am sure.

But even more important you will be seeing on the Garst family farming operations the most modern methods being used in the production and handling of corn as a crop, and among the most modern methods of swine raising and of chicken raising for egg purposes. You will get a very fine look at Midwest American agriculture.

You will not get to see beets nor potatoes. We do not raise either crop in this vicinity. But you will get to see a broad spectrum of Corn Belt agriculture that should give high interest to you. I hope and expect that the visit will not only prove to be interesting, but that it will inspire you all and give you a determination to have Polish agriculture advance rapidly and efficiently, and along every front.

I have written this letter simply to try to be helpful in the broad direction I think your recommendations should take. We will have ample time to discuss it at length during your visit here.

Very sincerely yours,
Roswell Garst

Hungary

March 14, 1958

To Minister Imre Dogei, Mr. Lazlo Fono, Mr. Bela Gonda,
Mr. Bela Molnar, Mr. Steve Sugar,
Ministry of Agriculture, Budapest
and to Mr. Sandor Rajki, Martonvasar, Hungary

Gentlemen:

The SS *Ardyk* carrying your Pioneer hybrid seed corn, and your single crossed hybrid seed corn, having left Norfolk March 10 should be at least nearing Europe. The exact amounts in bushels of the several Pioneer varieties are as follows:

Pioneer 371	1,600 bushels
Pioneer 347	7,903 bushels
Pioneer 344	7,406 bushels
Pioneer 329	12,000 bushels
Pioneer 325	2,892 bushels

Pioneer 371 is the earliest of the five hybrids. Pioneer 347, the next earliest; Pioneer 344 about like Pioneer 347; Pioneer 329 and Pioneer 325 may be a day or two later than the Pioneer 344 and Pioneer 347. . . .

In the case of each of these varieties, the male and female can be planted at exactly the same time and the flower on the male rows should be shedding pollen at the same time the silks on the female rows are emerging.

In our own case, we use a four-row planter and put the male corn in the outside box. That way we keep the male corn in one box and the female corn in the other three boxes and no changing of corn within the planter boxes is necessary in the planting of the field.

In the production of seed, we try not to have the population of the corn in the seed field too heavy. We try to plant a population of about twelve thousand or fourteen thousand kernels per acre, which lets us end up with a stand of ten thousand or twelve thousand plants per acre, which gives us nice size of seed ears. With proper fertilization and with decent moisture, such seed fields have averaged yields for us of above ninety bushels per acre on the female acres. I am sure you have had enough experience in the production of seed so that you will realize the necessity of careful planting and very careful detasseling. . . .

Because of the lack of complete cordiality between the governments of the United States and Hungary, it is still necessary to secure a special validation of American passports to travel to Hungary, where that is no longer true concerning Rumania, Yugoslavia, Czechoslovakia, Poland, or Russia. The State Department, however, seemed to be very willing to validate my passport last December, and I think they do validate passports for Americans who have legitimate business activities or a legitimate reason for visiting Hungary. And likewise, I believe the American government will be perfectly willing to issue visas for a small delegation from Hungary to come to the United States to study things that are of importance to your country. However, I think it would be wise to plan the visit rather well in advance.

You can see a great many things at most any time of year. For instance, in early May, you can see corn planting going on, or fertilization methods, the hog and chicken operation, and matters that would be of high interest to you.

Or, in July you could see the corn being detasseled. In June you could see it being cultivated. It depends upon what you want to see most and when would be your most convenient time. . . .

I almost forgot to tell you that we sent you, with our compliments, twenty bags of Garst & Thomas hybrid grain sorghum seed for experimental purposes. The bags weigh fifty pounds each. Because of a bad harvesting season, the germination on it is only 70 percent so it should be planted at the rate of eight pounds per acre. Just drill it with a corn planter in forty-inch rows or in closer rows if you desire. . . .

Sincerely yours,
Roswell Garst

February 15, 1960

To Mr. Lazlo Fono,[1]
Budapest, Hungary

Dear Mr. and Mrs. Fono,

. . . . I have been disappointed that no delegation of agriculturalists from Hungary visited the United States this year. The American agricultural techniques are letting the United States make terrific expansions in the production of agricultural products and always with less and less labor. The techniques which we have developed are worthy of imitation—and ought to be imitated much more rapidly than they are being imitated.

Particularly those areas in the Danube valley which lie in eastern Czechoslovakia, Hungary, Rumania, Bulgaria, and Yugoslavia where you have fertile land and good cultivating conditions could double and triple their production by the use of irrigation, heavy fertilization, insecticides, and herbicides. And you could double and triple the production with half or one third as much labor as you now use with proper mechanization.

Hungary has high mechanical ability so you could make your own machinery—you only need samples—you only need to see the opportunity. Western Europe does not have enough good land and warm weather for their ever-increasing population. They are going to have to depend upon the Danube Valley for ever-increasing amounts of grain and meat and eggs and poultry. Even the people who live in the cities in Hungary are paying too high a price for their food—far higher than they would need to pay if your production were up. So the agricultural opportunity of Hungary is simply terrific.

I thought last spring after our long visit with the assistant minister of agriculture and my cordial invitation for him to visit with us that my invitation would be accepted. This did not prove to be true.

Not only was I hopeful that he would come—I was hopeful that you would accompany him. I still am hopeful of being able to show you the many things which I think would work to the advantage of Hungarian agriculture. It seems to me that you and Mr. Rajki and the assistant minister of agriculture all three should come over.

Why don't you take the matter up with your government and see what can be accomplished? I have invited people from Rumania and we expect

---

1. A former aristocrat and expert in garden seeds who retained a position in the post–World War II agricultural establishment.

to have a continuation of visitors from the Soviet Union—and it seems that Hungary should have at least as much interest as those two countries. . . .

Sincerely yours,
Roswell Garst

October 18, 1960

To Mr. Stephen Tompe, Deputy Minister of Agriculture
and Mr. Gabriel Petohazi, Deputy Minister of
Agriculture, Budapest, Hungary

Gentlemen:

First I want you to know that Mr. Fono and Mr. [Sandor] Rajki arrived last week, October 13; that we have enjoyed their visit more than you can know; that we have shown them the Midwest, the great grain producing area of the United States at the peak of the harvest season; and that our own disappointment is that you two gentlemen could not accompany them this fall. . . .

We feel that the fastest and surest profit will come to your country through the purchase of seeds. I have told Mr. Rajki that I think your country should purchase a minimum of ninety tons of Pioneer hybrid seed corn and a minimum of ten tons of hybrid grain sorghum seed. I am especially anxious that you purchase both the seed corn and the hybrid grain sorghum seed of several varieties as to maturity for checking within your own country. If you wish to buy foundation stock of the corn, it will be available to you in the form of single crosses but not in the form of inbreds, and I would discourage the purchase of single crosses until you have decided which of the several varieties you like the best, and I think it would be wise of you to compare them not only with your own Hungarian hybrids but also with any other hybrid corns from any other source that you may be using next year.

When it comes to the foundation stock of the egg-laying type of chicken, I am completely certain that these should be bought from the Pioneer Hi-Bred Corn and Chicken Company of Des Moines—chickens sold under the trade name of Hy-Line. They are superior in breeding and in egg-laying production to any hybrid chickens anywhere and also in livability and feed efficiency. They are the optimum in productive ability and efficiency. . . .

The Soviet Union did buy some of these Hy-Line chickens in the spring of 1956 and they have told us that their egg production from them was forty eggs more per year than from their own best breeds. If you get forty eggs extra per bird—and only fifty birds from each foundation stock pullet—you would get two thousand extra eggs for every two dollar investment or thereabouts—a literally fantastic profit, a profit not of 100 percent, a profit of more than 1000 percent—and you will get that with higher egg quality. . . . Mr. Chrystal pointed out that you are already producing reasonably good hybrid seed corn—and we have suggested only small amounts of hybrid corn. Mr. Chrystal thinks your hogs are quite good—and that cattle are slow and expensive. . . .

Mr. Rajki and Mr. Fono will leave here this afternoon to spend some time with Mr. George Finley of the Finco Company and to see some of the modern egg-laying plants so that they can study the necessary equipment for the most modern feeding and housing and perhaps even incubation. Mr. Finley will make a similar report about that situation.

And now we come to the last major recommendation and that is for some modern corn growing machinery. I am listing it on a separate page with full explanatory remarks and approximately the correct prices figured on an F.O.B. American port basis crated for shipment. We exported—that is, Mr. Grettenberg of the Grettenberg Implement Company did—a large part of these machines to the Soviet Union last spring, and we are using the prices as of last spring, which may be 5 percent lower than at the present time. . . .

In relationship to the machinery, we think it would be wise to have an American technician arrive in Hungary at the same time the machinery arrived there to not only help assemble the machinery—but to instruct your technicians and to service the machinery for a period of anywhere from two months to six months at your decision. We also think that the machinery should be concentrated on one large state-owned farm so that one American technician could have the repair parts and all of the machinery under his direct supervision—and a place where your technicians could go and study the things that are being done. The distances are not so great in your country but that all of your technicians and state farm managers could visit that one farm, and we believe it is preferable. . . .

We realize—Mr. Chrystal, my son David, and myself—that the demands upon the Hungarian government for capital expansion are very great for industry, for transportation, for housing, for every type of investment as well as for agriculture. However, you are a nation of ten million people or thereabouts—and we think agriculture simply cannot go forward in Hungary as rapidly as it should go forward without selecting the most profitable parts for you to enlarge. An expenditure of five cents per person, which would be $500,000 per year, would let you start on the most important phases. We believe, actually, that it should be on an annual basis—with the most profitable things coming first, and that you should plan for the future years about such an expenditure as an absolute minimum.

You are going to certainly want to add the production of chickens for meat purposes; you are going to want to add a division of cattle for beef and dairy improvement, improved slaughtering and refrigeration methods; better fruit quality and harvesting and control; and permanent feed mixing installations for the preparation of finer feeds for your chickens and your hogs.

We actually believe that it is almost essential even this first year to buy enough feed additives to get you maximum production of eggs. In the United States, we mix about one hundred pounds per ton of feed in the form of the finest selections of minerals, antibiotics, vitamins, etc., to the normal mixture of grains and proteins. These one hundred pound additions are known as feed additives, and they have an extremely high return in the form of additional eggs. . . .

I can only say in closing this letter that Mr. Fono, Mr. Rajki, Mr. Chrystal, my son David, and I have spent five days discussing the most effective possible methods of helping Hungarian agriculture—and we have tried to limit our recommendations for immediate purchases to those things which would bring you much the greatest profit per dollar invested.

Very sincerely yours,
Roswell Garst

August 9, 1962

To Mr. Lazlo Fono,
c/o Ministry of Agriculture
Budapest, Hungary

Dear Mr. Fono,
. . . . Neither the agriculture of the Soviet Union nor the agriculture of Hungary, Rumania, nor Bulgaria—I class all three governments together in this respect—has so far been willing to put enough capital investment in agriculture to really expand agricultural production and at the same time reduce agricultural labor requirements.

I know that following the war you had to rebuild your cities, your transportation systems, your factories, your electric systems, that of necessity you made your first capital investments in industry and housing other than agriculture, and I am not critical of this judgment.

But because of the lack of capital, your agricultural production has not risen like the agricultural production of the United States—*which it could easily do!* And you have not reduced the labor requirements of agriculture—which you could easily do.

In the United States, the government is spending some five billion dollars a year trying to curtail agricultural production. That is out of a national budget of eighty or ninety billion. If we could spend more than 5 percent of our national budget trying to curtail agricultural production, certainly the governments of Eastern Europe should be willing to spend 5 percent to 10 percent of their national budget trying to build up agricultural production.

And this 5 or 10 percent of your national budget that ought to be spent on increasing agricultural production and on lowering materially the agricultural labor requirements ought to mostly come from your own resources. You have skilled mechanics, excellent factory facilities, an intelligent industrial complex—you could build the things you need and ought to do so.

But it seems to me that because we have done such a magnificent job in agricultural production and in agricultural machinery and in designing labor-saving agricultural equipment, you ought to take advantage of the opportunity that exists to thoroughly study everything we have before you start building it yourselves in quantity.

I know dollars are short. I know foreign exchange is short. But I know also that as individuals or as nations, we simply have to do the most important things—and that finally we are able to do them.

Your government, in my sincere opinion, should send a small delegation to the United States and purchase several complete sets of American farm machinery of the most modern design. It may interest you to know that the Garst Family here in Coon Rapids are now using machinery so modern that practically none of it was available in 1955.

Our new tractors are diesels—much more powerful, much longer-lasting, much more efficient. Our new planters are eight-row planters instead of four-row planters, eight-row cultivators instead of four-row cultivators; our new corn pickers are bigger and stronger; our equipment for gathering the corn stalks after the corn has been picked is very greatly improved. The new and modern combines are equipped so that the chaff can be saved by itself or the chaff and the straw can be saved together for the wintering of cattle. We have greatly improved yields by the use of soil insecticides—and also by the use of herbicides.

Our own corn crop this year will be better than the magnificent crop that you did see two years ago. When Mrs. Garst and I were married over forty years ago, it used to take a half an hour of a man's time to produce a bushel of corn. Now the average for the state of Iowa is four minutes of man time for a bushel of corn—and the Garst family are actually raising a bushel of corn in two minutes of man time.

When Mrs. Garst and I were married, it took one-third of the total work force of the United States for agriculture. It now takes less than 10 percent for agricultural production—and we are completely certain that 5 percent of the nation's population will soon be all that is necessary to produce surplus food above the finest diet the world has ever seen.

Hungary is a fine, intelligent group of people. Your country contains a very great deal of the world's finest land. You have an opportunity to expand agricultural production exactly as we have expanded agricultural production. You need irrigation, of course, but large areas in the United States where crops were poor formerly because the area was without sufficient natural rainfall are now among the most productive areas in the United States because we do have irrigation.

You need fertilizer, of course, but fertilizer is largely a matter of the nitrogen out of the air being fixed mechanically, the only three ingredients being air, water, and a little natural gas with plenty of power. You can pipe up the natural gas from the Rumanian fields if you have to. Then you need all kinds of small tools which you could readily make.

I will never forget the first delegation that came from Hungary[1]—Rajki, [Janos] Keseru and [Mikos] Csillag. Keseru was, at the time, managing a very large state-owned farm. He told us he was using "twenty men and forty oxen" to water the livestock on the farm that winter. We were using a two-horsepower electric motor on a little pressure water system and watering far more cattle and hogs than was Keseru. You ought to take one factory and have it start making simple little water systems. . . .

I do believe that you should interpret this letter immediately and discuss the matter with the minister of agriculture and his deputies—I would even be hopeful that the minister of agriculture could discuss the whole matter with the Cabinet and secure an approval on whatever amount of dollars the whole Cabinet feels justified in spending before the delegation comes to the United States so that purchases could be made while the delegation is in the United States.

We folks here at Coon Rapids will be just as helpful as possible in

1. December 1955 to January 1956.

advising with a delegation and picking out those items for purchase which we think would be of greatest benefit to the Hungarian nation. . . .

Sincerely,

Roswell Garst

October 21, 1969

To Mr. Istvan Gergely, Deputy Minister of Agriculture

Mr. Lajos Lada, Head of Agricultural and Food
Industry Department in Ministry of Foreign Trade

Mr. Lajos Toth, Deputy Head of International
Relations Department in Ministry of Agriculture

Mr. Andras Klenczner, Managing Director of
National Center of State Farms

Dr. Robert Burgert, Director of State Farm, Babolna

Mr. Peter Sarkozy, University Lecturer

Mr. Balint Szaloczy, Agricultural Attaché and
Second Secretary, Embassy of Hungary

Gentlemen:

First, I wish you all to know how proud and pleased we were to have you visit Coon Rapids—and the Garst farming operation and the Garst family. . . .

Your delegation purchased some Pioneer hybrid seed corn—some seed corn drying and sizing equipment—some farm machinery—some small water pressure systems, etc. . . .

There are other things that I believe you should be purchasing in the United States—because we have had enough experience to have eliminated the faults. Such as:

Fertilizer equipment for transportation, storage, distribution. As you saw while you were here, the Garst Company retails fertilizer in the Coon Rapids area. The Garst Company is a partnership of Stephen and David Garst, my two sons.

They have a bulk dry blending plant. They have a bulk liquid blending plant. They have little trailers to haul either dry or liquid fertilizer from town to the fields and spread it. These trailers are hauled behind any tractor.

Fifteen years ago all fertilizer was shipped and handled in paper bags.

Now almost all of it is handled in bulk. This eliminates the cost of sacks, the work of lifting; it is far more economical and more profitable.

Hungary should, *without fail*, send a very small—two or three men—delegation over and construct some dry blend plants and buy some dry blend equipment and perhaps even a liquid plant or two.

We in the USA have wasted lots of time and money before we learned that bulk handling was far more economical than bagged fertilizer.

Then farm machinery has been improved very greatly. Tractors are more powerful—corn planters are now eight-row or twelve-row at a time. Corn combines are rapidly replacing corn pickers.

To show you how completely changed American agriculture is, let me point out that up to 1955 the national average corn yield was thirty-nine bushels per acre—now it is seventy-eight bushels per acre. In short, yields have doubled from an already decent yield to an extraordinarily high yield by the use of fertilizer, insecticides, herbicides, better hybrid varieties, better mechanization, and therefore more timely planting, harvesting, etc.

When you were here I showed you a fine large warehouse for the storage of grain, or cobs, or anything, that had been constructed entirely from cottonwood lumber.

If you had several small sawmills and some equipment to plane the boards, and you used the laminated type of construction, you can use cottonwood, elm, maple—almost any type of trees—because by laminating you do not need long lengths of boards.

You need more warehouses and they are economical to build.

And you just have to irrigate to take full advantage of fertilizer. Irrigation is not very profitable without fertilizer. And fertilizer brings limited response without water. It is when fertilizer and irrigation are combined that production becomes phenomenal. . . .

The development of "walk around" irrigation has increased the American potential [for] agricultural production phenomenally. It can and should do the same for Hungary.

Your delegation talked about the "shortage of dollars." Every American farmer has that same problem—"shortage of dollars." Every American knows the way to cure the "shortage of dollars" is not to raise less—the way to cure the problem is to raise more.

In the United States the United States government subsidizes agriculture by about $4 billion a year. We have 200 million people. So that is about $20 per person.

It seems to me the Hungarian government should be willing to spend $1 per Hungarian citizen per year for a few years to stimulate more production in agriculture. My best judgment is that such an expenditure would return extraordinarily large returns in the increased export of foods of all kinds.

As I showed you while you were here, the USA has very good country roads—everywhere. They are not expensive to build because we have the world's best road grading and road maintenance machinery.

I know you were an *agricultural* delegation—not a road building delegation. But—

Better than many people I know how extremely important good roads are to agriculture. And I know that as agriculture "goes modern" good roads are not a luxury—but are absolutely essential.

I think Hungary—either the Department of Agriculture or the Department of Roads—ought to buy a couple of sets of road building and road maintenance equipment. . . .

It seems to me that improving Hungarian agriculture is not a matter of taking from the city residents to help rural residents—it is a matter of helping rural residents so they can provide more and better food to city residents at less cost to the city residents and do so with less labor.

You members of the agricultural delegation had a chance to see the very high production—not only per acre—but per man hour.

Just think of this one thing! When Mrs. Garst and I were married nearly forty-eight years ago, it took thirty minutes of man time to produce and harvest a bushel of corn. Now in the United States the average is between four and five minutes per bushel of corn—and quite a few farmers can raise and harvest a bushel of corn in less than three minutes of man time. . . .

Most sincerely,
Roswell Garst

November 14, 1969

From Sandor Rajki
Agricultural Research Institute
Hungarian Academy of Sciences
Martonvasar, Hungary

Dear Mr. Garst,
. . . I wonder if you have heard about that after long years Hungary is again exporting wheat: in the current year, a total of 600.000 metric tons

will be exported by Hungary to the Lebanon, Czechoslovakia, Britain, Switzerland and Italy. In recent years, Hungary has not needed bread grain imports any more, and by this year a considerable export volume could be effected. The latter is due to the highest national average yield on record in the country, it is about 2800 kg/per hectare.

As to the corn yield these days we finish to harvest—similarly to wheat—the highest national average grain yield on record estimated about 3500 kg/per hectare.

It seems to me that this is the time you should visit again us, Mr. Garst.

With best regards.

Very sincerely,
Dr. Sandor Rajki

P.S. Erna joins me in sending warmest family to family greetings.

September 23, 1971

To the Hungarian Delegation
Visiting Garst & Thomas Sept. 23, 1971

Gentlemen:

Garst and Thomas are pleased and proud to have you visit us. It will have been sixteen years ago in early November since I visited Hungary the first time. That was November of 1955.

Even at that time, Hungary knew about hybrid seed corn—and was starting to produce some small amounts of hybrid seed corn.

But Hungary was behind the United States in modern machinery, in the use of fertilizer, insecticides, herbicides, and many other new methods of raising "more and better food with less labor."

You have doubled corn yields and materially reduced labor requirements in the last sixteen years. You have expanded both poultry and swine production rather phenomenally. Your agriculture has made very great progress.

There is, of course, more progress that you can make. Let me suggest a few things you should do.

1. Because the United States produces about one-half of the world's total corn crop—125 million metric tons in 1971—it surely

would seem logical for you to compare some of our most modern Pioneer hybrid varieties with your own. . . .

2. In the USA in any area where the annual precipitation is less than six hundred millimeters, grain sorghums are grown on as many hectares as corn. In areas where precipitation is less than five hundred millimeters, grain sorghum is planted on far more acres than corn. I urge you to try grain sorghums widely in Hungary. Plant them in 750 millimeter rows; plant them with modern plates. Fertilize the grain sorghums.

Harvest them at 30 percent moisture—grind and ensile the ground high moisture sorghum.[1]

High moisture ground grain sorghum is 18 percent more efficient than is ground dry grain sorghum. It can be fed to swine and poultry as well as cattle.

Grain sorghums can wait longer for rain without being damaged than maize and for that reason are more useful in areas of undependable precipitation.

3. You simply must use the urea form of protein for all of your ruminants—all your cattle, sheep, or goats.

Ruminants can use the urea form of protein as well as they can use soybean meal, sunflower meal, rape seed meal, or fish meal. Use urea for all of the protein of ruminants so you can have more vegetable protein for the poultry and swine.

4. There is increasing evidence that minimum tillage is a way to save moisture, to save man hours, to save tractor hours and to increase crop yields. It seems important to me to urge you to switch from discing, plowing, discing, planting, harrowing, and cultivating three times as we used to do to discing, planting, cultivating once, and harvesting.

Every bit of experience we have had on the Garst farming operation stimulates our belief that minimum tillage is gaining everywhere in the USA.

5. With unlimited and very economical protein for ruminants available in the form of urea the use of corn stalks, corn cobs, grain sorghum stubble for the feeding of beef-brood cows or pregnant ewes is almost unlimited. A pound of protein in the form of soybean meal is more than 10 cents. A pound of protein equivalent in the form of urea

1. That is high-moisture sorghum that has been ground up.

is 1.25 cents. You must learn to use urea for *all* of the purchased protein. . . .

You have my permission and my encouragement to interpret the letter and the literature into Hungarian and use them in any way that will be of benefit to Hungarian agriculture.

Very sincerely yours,
Roswell Garst

December 21, 1972

To Ambassador Karoly Szabo
Counselor Peter Fulop
Embassy of the Hungarian People's Republic
2437-15th St., N.W.
Washington, D.C. 20008

Gentlemen:

My nephew, Mr. John Chrystal, and I have recently returned from a rather short and hurried trip through the Soviet Union, Rumania, and Hungary. The trip was too short—undesirably short—but even a short trip was enough to let us see the almost unbelievable progress Hungary has made since we were there in late May of 1963.

In two weeks, the year 1973 will have arrived. We both send our greetings to you gentlemen—and to the country you represent—and our hopes that the year of 1973 will be one of peace in the world and a year when expenditures for armaments in every country in the world will be less than they have been in the past. . . .

Primarily Mr. Chrystal and I were encouraging the introduction of hybrid grain and pasture sorghums into Eastern Europe. The Soviet Union had used sample amounts in the year 1972 and were so impressed with the results that they have ordered seven hundred tons of grain sorghum seed for 1973 planting and will send a group of engineers over to build [sic] a drying and treating plant for producing their own seed.

The Rumanians have tentatively ordered one hundred tons and also some farm machinery.

We were only able to spend a day and a half in Hungary but it was a wonderful day and a half.

We had phoned Mr. Sandor Rajki, director of the Martonvasar Agricultural Research Institute, from Bucharest, and he had reserved rooms for us. He met us at the airport, and we had a long, long visit. In the morning, we had a meeting with Mr. Rajki and the deputy minister of agriculture and then drove down to Martonvasar for a look at the wonderful new building for the controlled conditions for year-round research into all kinds of plants.

We believe that the Hungarian Department of Agriculture will probably purchase a minimum of fifty tons of hybrid grain sorghum seed—and some forage and pasture sorghum seed as well.

While very great progress has been made—and is being made—in the agriculture of Hungary, I do believe that the purchase of some of the most modern farm machinery seems to me to be very worthwhile.

In the United States farm labor is so expensive that we have been forced to go more and more to labor saving equipment.

Because the United States produces more than half of the world's corn and about 75 percent of the world's soybeans and most of the grain sorghum that gets into international commerce, we have been under pressure for many, many years to constantly improve the efficiency of our agricultural machines.

I urge you to encourage your minister of agriculture to feel free to send several small delegations to the USA next spring and summer. I suggest two machinery men and a *really competent* interpreter, as the right sized delegation. That way, I can take them in a car comfortably and show them what machinery we use—and what is in general use. . . .

Very sincerely yours,
Roswell Garst

cc: Sandor Rajki

September 30, 1974

To Dr. Sandor Rajki, DSc
Agricultural Research Institute
2462 Hungarian Academy of Sciences
Martonvasar, Hungary

Dear Sandor:

Dr. Andor Balint is here—in fact, has been here since Friday, September 27—and is leaving today.

I have enjoyed showing him everything I thought would be helpful to the agriculture of Hungary, but—

I believe the *minister of agriculture* of not only Hungary but of every country should visit the USA. I will outline the reasons!

You, Sandor, have one part of agriculture that you have mastered, "hybrid wheat"; Andor Balint has another part he is interested in—that is, more protein and better protein in corn!

Hungary has specialists in everything from strawberries to watermelons to potatoes to ducks to geese! So does every other country!

"But again"——

You refuse to look at waste that is obvious! For instance, you put up corn silage in piles on top of the ground. It decays in from the top—from both sides and both ends. The decay must affect 20 percent as a minimum. A trench silo, with a cement floor and cement tilt-up side walls is very inexpensive and you can have almost no loss.

We have machinery for harvesting the silage that is very efficient, other machinery for feeding it that is very efficient. It is machinery that Hungary could manufacture yourself in your own country—but you need samples of it.

Iowa State University at Ames has finally agreed that to waste corn stalks and put up hay on the same farm the same year is ridiculous—that corn stalks with proper supplements are as good for cattle as is alfalfa hay. And they now have the right machinery to put [up] the corn stalks. . . . [1]

"But again"——

The world is short of food. Think of this! On the Chicago Board of Trade corn is selling now for $3.50 per bushel for December 1975 delivery, and is 6.7 cents per pound—$134 per ton.

Not according to me, but according to Iowa State University, Michigan State, and every university, one acre of corn silage fed to cattle is almost two times as valuable as shelled corn from the next acre.

I hope you put this letter into Hungarian language and make copies of it for your Bureau of Ministries.

I urge you and them not to give the matter any publicity but to send a small delegation over here—a delegation authorized to spend $100,000 or $200,000 buying the most advanced machinery for harvesting corn—and corn stalks separately—the proper methods for ensilage, etc.

I want a report from you on what you think of this idea.

Sincerely,
Roswell Garst

1. That is, convert the cornstalks to silage.

October 21, 1974

From Sandor Rajki
Agricultural Research Institute
Hungarian Academy of Sciences
Martonvasar, Hungary

Dear Mr. Garst:

I have informed Mr. Imre Dimeny, the Minister of Agriculture and Food Industry, both in writing and yesterday verbally of the suggestions you made in your letter of September 30th. These suggestions are at present under consideration, though due to the cancellation of the long-planned visit to Hungary by Mr [Earl] Butz, the US Minister of Agriculture, I think it unlikely that a ministerial visit to the USA will take place in the foreseeable future.

You are already well acquainted with my personal opinion on the suggestions you made concerning the utilisation of maize stalks, a large proportion of which go to waste here, and concerning the complete exploitation of all the possibilities of maize. We have discussed this a great deal with you and your family. You can be certain that your suggestions and ideas on the development of Hungarian agriculture were not just voices crying in the wilderness. They were forwarded to the appropriate Hungarian authorities, even in those cases when the ideas, though in my opinion truly efficient, were not put into practice on the Hungarian side.

I have recently received a letter from Dr Balint, in which he assesses his three-day visit to the Garst farm as the greatest professional experience of his American tour. I was sure it would be so, knowing the human and professional worth of you and Mrs Garst, and of all the Garst family. It was particularly gratifying to learn from Dr Balint's letter that you and Mrs Garst spoke of us, Erna and Sandor [Jr] too, with such affection. Recently, I had occasion to hear Sandor [Jr] speaking to an American guest and I was amazed at the ease and accuracy with which he speaks English. Please be assured that Erna and I really appreciate not only this, but also the experience that Sandor obtained, with your friendly assistance, during his 2-month visit to America.

Here in Hungary the rain, which has not let up for a month, is causing serious concern to agriculture, as regards both harvesting the crop and completing the autumn sowing. But we have had an excellent year, with record cereal crops. We have heard about the drought which hit certain areas of the USA during the growing season and about the early autumn frost. Knowing American efficiency and business ability, however, we are sure that these difficulties will be overcome.

Marvelous things are underway in our phytotron.[1] It seems that adequate variability can be induced in the phytotron too, e.g. the transformation of [Norman] Borlaug's Mexican spring wheats into hardy winter wheats, by means of appropriate cultivation, which opens up a seemingly boundless perspective both for biological and agrobiological research and, through this, for practical agriculture. If only we could live for at least 100 years, so as to give our dreams time to materialise. What a pleasure it would be to discuss this some time with you, Mrs Garst and the whole Garst family, in the sort of atmosphere that was created, for instance, that evening we spent together in the summer house on Lake Panorama,[2] after I came back from Mexico. At home in Martonvasar, as you know, it is in Erna's company that I am able to weave my scientific-professional dreams.

I do not know who I have to thank for the invitation to lecture at the Baghdad University, since I have no personal acquaintance there. All expenses will be paid, including the airfare. Such an invitation is not to be refused. I plan to fly to Baghdad on November 10th in order to give several lectures on genetics, and I should like to make use of my 10 days' stay in Iraq to become acquainted with the monuments and ruins of one of the cradles of human civilisation too.

We are all well.

We all, Erna and the children, Sandor and Katalin too, send you all our very best wishes.

Sincerely,

Sandor Rajki

June 5, 1975.

From Sandor Rajki

Dear Mr Garst,

Many thanks for the interesting literature you sent me earlier, "Insulated Hungary feels pinch of world problems", for your letter May 15th and for the booklet "Profit in Stalks Handbook". . . .

It would be very interesting to see the new machine on the Garst farm, which you wrote about in your letter and in the "Profit in Stalks

---

1. The phytotron is the laboratory at Martonvasar in which artificially controlled growing environments are created in order to modify or change wheat varieties, as Rajki describes.
2. Former summer home of David and Jo Garst.

Handbook", which I found exceedingly interesting and valuable. I told Dr Dimeny, the Minister of Agriculture and Food Industry, about your letter and the main points of the book, and I would be quite prepared to accompany him to Iowa in the autumn in response to your kind invitation. There is, however, a new question mark in connection with Dr Dimeny's possible trip to Iowa. There will be parliamentary elections here in the middle of June, so the government will also be reshuffled. Rumour has it that after 8 years in office Dr Dimeny may be replaced by a new minister. This is one reason why I delayed informing him of your letter and book.

The other reason was my trip to Cuba, from which I have recently returned. Fidel Castro invited me when he visited Martonvasar in 1972, but I have only just got around to accepting this really flattering gesture. It was a marvellous trip. How good it would be to discuss it with you and Mrs Garst and the Garst family, here at Martonvasar or in Coon Rapids.

Erna, Sandor and Katalin join me in sending our regards.

Sincerely,
Sandor Rajki

January 7, 1976.

From Sandor and Erna Rajki

Dear Mrs Garst:

We assumed from the letter you wrote together with Mr Garst on November 12th that you would be travelling south shortly afterwards. This is why we replied to David instead of to you and have not yet congratulated you on your 80th birthday. But as the Hungarian proverb says, good wishes are always welcome.

One of the great experiences of the past 20 years of our life was becoming acquainted with the most developed agriculture of our time, that of America. Due to a special providence, in the form of Mr Garst's visit to Martonvasar 20 years ago, we got to know the American agriculture chiefly on the Garst farm, and there is not a more efficient farm in the whole wide world.

The kindness and affection with which you overwhelmed us helped to make our stay on the Garst farm unforgettable. We hope you will not think it pretentious of us to say that we have found you to be a worthy companion for Mr Garst.

We wish you every happiness, and above all good health, on your 80th

birthday, and hope that you and Mr Garst will live to celebrate many more
happy birthdays in good health and strength.

Sincerely,
Sandor and Erna Rajki

September 23, 1977

To Mr. Sandor Rajki
Agricultural Research Institute
2462 Hungarian Academy of Sciences
Martonvasar, Hungary

Dear Sandor:

. . . This is the busy season for Steve, Dave, John Chrystal, and the
Garst & Thomas Organization, in fact, everyone but me.

We are harvesting the largest acreage of seed corn in history. We
planted the largest acreage. In the Coon Rapids area we had the most
severe drought in history, but we only planted about 10 percent of the
acreage where crops are poor. The other 90 percent is yielding 10 percent
above expectations, so we are okay.

You will be interested to know that we have four or five Hungarians
coming September 28, 1977, who are visiting Pioneer Hi-Bred International, Inc. in Des Moines. They are only going to be here for a few hours so
I cannot show them as much as I would like. . . .

One of the things I want to insist that you do in Hungary is to put on a
demonstration of how to use the cellulose of all the corn you raise, and that
you use the 10 percent urea and 90 percent molasses self fed as the protein.

It seems to me that if you have some young Hungarian men who can
speak English, it might be worthwhile to send over several of them for the
summer.

The world population is gaining at the rate of about 80 million a year.
Every twelve years we add about another billion people to world population.

So we all need to learn to "waste not, want not."

Sincerely,
Roswell Garst

1.

1. Wheat being combined in the USSR with primitive machinery about the time of Garst's first visit in 1955. During his years in power, Nikita Khrushchev strove to expand and improve the mechanization of Soviet agriculture, an area neglected during the Stalin years. (Courtesy William Robert Parks and Ellen Sorge Parks Library, Iowa State University.)

2. Russian women winnowing wheat. In the fifties, Soviet agriculture was extremely labor-intensive. As a result of the enormous casualties suffered by the USSR in World War II, women constituted the bulk of the agricultural labor force in the postwar decade. (Courtesy William Robert Parks and Ellen Sorge Parks Library, Iowa State University.)

2.

3.

3. Agriculture was extremely primitive in postwar Rumania. Here horses and wagons bring their loads to be threshed, in a scene reminiscent of US farming much earlier in the century. (Courtesy Harold Smouse.)
4. A Rumanian peasant carries a basketfull of newly picked corn from a field in 1956. The cornstalks in the rows behind him have been cut down by women. (Courtesy Harold Smouse.)

4.

5. Stephen Garst and Harold Smouse in 1956 demonstrating the operation of a Rumanian well from which a herd of livestock was watered. Roswell Garst describes this commonly used type of well in the first letter of this edition (2 November 1955). (Courtesy Harold Smouse.)

6. Garst, at the head of the table, about to enjoy a fish dinner at a peasant's house in the Danube delta in the autumn of 1955. Geza Schutz is standing in front of the window; geneticist Grigor Obrejanu is seated at the far right. Officials at the American legation in Bucharest, highly restricted in their movements, were amazed that Garst had traveled so widely in Rumania. (Courtesy Elizabeth Garst.)

5.

7. Rumanians were the first Eastern Europeans to visit Coon Rapids shortly after Garst's initial trip in 1955. Here, Virgil Gligor, Garst, Silviu Brucan, and Grigor Obrejanu view a cattle feedlot in December 1955. (Courtesy Elizabeth Garst.)

6.

7.

8. On 23 September 1959, the succession of visitors to the Garst farming operation reached its climax with the visit of Chairman Nikita Khrushchev. Here, Khrushchev (second from left), his son-in-law Adzhubei, Garst, and Henry Cabot Lodge view a cattle-feeding operation near Coon Rapids. (Courtesy Elizabeth Garst.)

9. The Khrushchev family entertaining Garst and John Chrystal at their country dacha in 1963. Seated (left to right): interpreter Marina Rytova, Chairman Khrushchev, Garst, Nina Khrushcheva. Standing (second from left): John Chrystal, Yelena Khrushcheva, Adzhubei. (Courtesy John Chrystal.)

8.

9.

10.

10. Fidel Castro, notebook and pencil in hand, tours the Hungarian Agricultural Research Institute at Martonvasar in 1972. His arm rests on the shoulder of Director Sandor Rajki, one of the earliest visitors to Coon Rapids in the fifties, and by the time of the picture, a leader in the modernization of Hungarian agriculture and a close friend of Garst. (Courtesy Sandor Rajki.)

11. Geneticist Ivan Khoroshilov, economist Alexander Tulupnikov, and former minister of agriculture Vladimir Matskevich, at a party in Moscow given by John Chrystal in 1983. Tulupnikov and Matskevich, who originated the Coon Rapids connection, are retired (as is Khoroshilov), but their successors maintain the USSR-Iowa relationship. (Courtesy John Chrystal.)

11.

12.

12. Twelve-row corn planter at Kuban Machine Testing Station, mid-1950s. (Courtesy William Robert Parks and Ellen Sorge Parks Library, Iowa State University.)

13. Silage cutter on Elicha Lenin Collective farm, mid-1950s. (Courtesy William Robert Parks and Ellen Sorge Parks Library, Iowa State University.)

13.

USSR, 1955–1959

Garst's immediate success with the Soviet delegation in Iowa was repeated during his first trip to the USSR and reached its climax in his meeting with Nikita Khrushchev. Orders were placed for hybrid seed corn and parent genetic stock in accordance with Khrushchev's determination to hybridize Soviet production and to create a hybrid seed corn enterprise as rapidly as possible. Garst entertained a Soviet delegation of experts in Coon Rapids, returned to the USSR in 1956, and, in addition to his lecturing, produced a pamphlet on the nature of Midwest farming practices that was translated into Russian, and later into Rumanian. After the hiatus in contact caused by the 1956 Hungarian uprising, Garst entertained further delegations and sold more seed corn. Khrushchev's desire to acquire the entire technology of the Garst operation was demonstrated by his insistence that two Soviet agriculturalists spend the summer of 1958 working in Coon Rapids, by his willingness to let Garst visit him again in spring 1959, and by his own visit to the Garst farm that autumn during his unprecedented American tour.

Garst's correspondence early in this period illustrates the professionalism of his approach to the challenge of this unique opportunity for sales: his preparation for the trip, his insistence on touring extensively in potential corn-growing areas of the USSR to see for himself the condition of Soviet agriculture, his willingness to disseminate his expertise and to follow through at home on Soviet requests. However, by 1959, the USSR was well on its way to producing its own seed corn, and Garst's relationship with Khrushchev had blossomed into that of principal Western adviser on agriculture. The emphasis in Garst's correspondence at that time shifts from business and technical details concerning hybrid seed corn to the need for the Soviet Union to follow the Western progression from hybridization to

mechanization and chemical fertilization. The hope that agricultural rap-
prochement could play an important role in establishing a new era of coexistence also became an important theme. At the Garst farm in 1959, asked by
a reporter whether Garst had put him into a good mood for his forthcoming
conference at Camp David with President Eisenhower, Khrushchev replied that Garst had been "creating that mood ever since we met in the
Crimea."

Thus in 1959 Garst had reached the pinnacle of his relationship with
Khrushchev. He had also reached the pinnacle of his frustration with the
rigid, slow-moving bureaucracy of the USSR. His correspondence at the
end of this period reveals irritation over the difficulties of executing business arrangements with Soviet authorities and his belief that his sweeping
recommendations for a new kind of revolution in the USSR were getting
lost in the corridors of the state ministries: "Your bureaucracy is as bad as
ours," Garst complained to Anastas Mikoyan in 1959. "Fellow Bolshevik!"
Mikoyan exclaimed, shaking Garst's hand in enthusiastic agreement.

September 9, 1955

To Mr. John W. Mathys
Northrup King Seed Co.
Minneapolis

Dear Jack,
    . . . I am leaving for Russia at the personal invitation of Mr. [Vladi-
mir] Matskevich, the head of the Russian delegation to the United States,
to visit the Moscow Fair and to study corn growing in southern Russia,
Rumania, Hungary, and Czechoslovakia. The State Department has
issued passports and the other countries have issued the necessary visas.
The State Department has further told me that they see no reason why they
would object to the Department of Commerce's issuing an export license,
and Commerce has said they see no reason why an export license should not
be issued if I get an order.

    I had lunch with the Russian Embassy and they understand that I am
coming to look over the situation and make recommendations for the purchase of hybrid seed corn. I frankly told the Russians that I did not consider
myself as a representative of the Pioneer group—that I considered myself a
representative of the top end of the American hybrid seed corn industry. I
specifically mentioned only Pioneer, DeKalb, and Northrup King in my

visits with the Russians—told them that there were a few others, reputable producers, and let it go at that.

After having made a good study of the thing—Jonathan Garst[1] sent me a marvelous set of tables showing temperatures and moistures of the several areas in direct comparison [with] cities in the United States such as North Platte, Huron, Minneapolis, and Des Moines; and after having visited with Dean [Vincent] Lambert of the University of Nebraska, who has just returned, and after having visited with Dr. Louis Michael, who spent twelve years in the area for the Department of Agriculture, during the thirties and forties, I know enough to know that the bulk of the corn they need would be classed as either early or extra early.

They will be able to use some medium maturity corn in southern Hungary and in extreme southwestern Russia around Odessa. They actually grow a little bit of cotton at Odessa. By medium maturity, I mean corns that will mature in south central and southern Iowa.

The bulk of the corn they will need that will be available will be corns that mature in northern Iowa and southern Minnesota, which I choose to call *early corns*. They will do excellently nearly anywhere in the southern Ukraine and in the Danube Valley countries.

Then they are growing just as much early maturing corn—I think I should say, extra early maturing corn—as they can, and by extra early, I mean stuff that would mature in the north part of Minnesota and in North Dakota. Where we have a relatively small potential area that demands extra early corn, they have a simply immense area that could use extra early corn.

They are going to plant 30 million acres of corn in the four countries mentioned that is in rather solid historic corn growing areas. It probably will take something like 3 million bushels of seed. They have no hybrid corn. There will not be anything like sufficient surplus of the correct maturities available from reputable firms that I would care to recommend. If they planted 30 percent or thereabouts with hybrid seed, it would take something like a million bushels—far more than I think is likely than the hybrid seed corn industry could possibly furnish them in the top varieties of surpluses. . . .

Now the Russians probably are familiar with the fact that a very great deal of hybrid seed corn was shipped to France, Greece, Italy, and Germany under the Marshall Plan and the ECA and that this corn was purchased at a relatively cheap price. I actually think some of it returned only $2.50 to $3.00 a bushel at the seaboard to the American producer. Of

1. Roswell Garst's brother, formerly an official in the USDA.

course, it was not worth anything like that in comparison with the really good hybrid strains. It was junk. It didn't have to germinate very well—nobody knew about its inheritance—it was poor junk.

I know that the Pioneer group and the DeKalb group and Northrup King are in no way interested in selling junk—and in no way interested in selling at a very cheap price. The big question in my mind is what the price should be. . . .

You understand, Jack, I don't know whether I can sell them a bushel of corn or not. I only think that I can come as near doing it as anyone else—and I think if they buy any to speak of at all, they will buy all that they can get their hands on that is really well adapted.

Every time Matskevich gave a list of things that impressed him he said that the thing that impressed him most was our hybrid corn. Every time Mr. Krushchev has said anything, he has said they want to raise more corn so they can have more meat. Dean Lambert told me that their corn had horribly poor breeding—great variation in height of ear, type of ear, great stalk weakness, etc., etc.

Actually, all he did was describe open pollinated corn. We used to have the same thing in the United States everywhere. And as you know, in the days when we had open pollinated corn, I was out running yield comparisons with every darned customer and "bellering and bawling" about the superiority of the hybrid over the open pollinated. At that game, I am a man with twenty-five years of experience.

So, actually, I have the darndest confidence you ever saw that I can go over there and sell them more corn than the Pioneer group and Northrup King and DeKalb group can furnish in the way of surplus of really well adapted varieties. I may fail—I have failed to make some sales in my time—but I never had more confidence. . . .

Incidentally, I think they ought to be producing their own hybrid seed—I think that is so logical that this market will only be good for a few years—perhaps as many as five. They will get some inbreds off the colleges or small growers, they will hire some geneticists, and it's only logical to think that they ought to produce their own seed corn. But I don't think they will be able to do so for three or four years in any quantity at all and probably not in sufficient quantities for five or six. . . .

Win, lose, or draw, I think I will have a great trip. I am taking with me a linguist from down at Janesville, Minnesota—one Geza Schutz, born in Budapest, a Ph.D. from the University of Geneva, a couple of years with the League of Nations, postgraduate work in Economics at Columbia, worked with the Labor Department in the United States for quite a few years in the late forties, was in fact, I think, the chief labor conciliator of the Minneapolis office—a grand guy and a grand musician and thoroughly familiar with everything I have ever done here at Coon Rapids. He doesn't just translate in a cold voice—he translates enthusiastically—and can wave his arms a little bit in real excitement.

I am not taking a camera, I am not going as a visiting fireman, I am just going as what I am, I am just going to do the thing that has always thrilled me the most—I am just going to try to sell a big order of very fine hybrid seed corn to plant in the place where open pollinated corn would otherwise be planted to bring profit and pleasure to the people who plant it. Some people like to fish, some people like to play golf, some people like one thing and some another. But I like to sell big orders of superior hybrid seed corn against open pollinated corn. That's why I think I am going to have a whale of a trip. . . .

Sincerely,
Roswell Garst

September 17, 1955

To Vladimir Matskevich
Minister of Agriculture

Dear Mr. Matskevich:

I wish you to know that while I expect to very much enjoy and benefit from visiting your fair in Moscow and while I wish particularly to examine the machinery that is used in connection with the growing of corn and the machinery that is used in the preparation of not only your corn grain, but your fodder, for the feeding of cattle and hogs, I think the principal way in which I can benefit your country and thereby, as I see it, benefit our own country is by a thorough examination of your corn at harvest time. I am therefore extremely anxious to go to the corn growing areas of the USSR during this harvesting period. I also wish to see the corn growing parts of Hungary, Rumania and Czechoslovakia for the same purpose—a study of your corn varieties and your cultural practices and soil types and weather conditions with the idea that I may be able to be helpful to your people and to learn something that will be helpful to me. I think this whole trip may well work out to the mutual advantage of the USSR and the United States.

However, after having been in the hybrid seed corn business since its very start, I feel sure that the greatest contribution that can be made in an immediate fashion would be the sale of well adapted, early maturing, and extra early maturing hybrid corns from the top producers in the United States to the USSR and perhaps to the governments of the Danube Valley countries mentioned.

I realize, of course, that over a long period of years your country and the Danube Valley countries should produce your own hybrid seed corn. It is not logical nor sensible for you to plan upon purchasing hybrid seed corn from America, except during the relatively short period of time which would be required for you to get your seed production facilities arranged. But certainly for the 1956 planting and most likely for at least a couple of years after that time, and perhaps for as much as perhaps five years after that time the hybrid seed corn industry in the United States can be helpful.

Ever after you invited me to come to Moscow to see the fair, I have been concentrating on these thoughts. I have looked up the temperatures and rainfall and soil conditions of your area and of the Danube Valley countries and compared them directly with areas in our country. I have interviewed people who are familiar with the corn growing areas of the Danube Valley and of your own country, and I am completely certain that if great care is exercised from a maturity standpoint, reasonably large quantities of very superior strains of American hybrids would be available, although not in as large quantities as you perhaps would like.

For your information the two largest producing companies in the United States are the DeKalb Hybrid Corn Company of DeKalb, Illinois, and the Pioneer Hi-Bred Corn Company of Des Moines. Garst & Thomas are associated with the Pioneer Hi-Bred Corn Company of Des Moines. The Pioneer Hi-Bred Corn Company of Indiana and the Pioneer Hi-Bred Corn Company of Illinois are in this same group.

The DeKalb company and the Pioneer group each produce and market domestically in the United States in the neighborhood of two million bushels each—for a total of approximately four million bushels. The great bulk of the corn that is sold by both of these companies is in the northern part of the country, where early and extra early corns are very widely used. There is another very reliable seed company in the area—The Northrup-King Seed Company of Minneapolis. They are very high quality producers and have a good deal of early and extra early corn.

Now after studying the matter over, I came to the conclusion that you could very profitably use much larger quantities of corn than the Pioneer corn that Garst & Thomas would have available—or that would be available through the whole Pioneer organization—and I came to the conclusion that it would be simpler and easier if I just came over as a representative, not

alone of the Pioneer group, but as a representative of the really top producers of hybrid corn in the United States. As a result I visited with Mr.
Tom Roberts of the DeKalb Hybrid Corn Company of DeKalb, Illinois,
whom you also invited to come to Moscow. And I visited with the Northrup-King company at Minneapolis.

Mr. Roberts' health is not the best, and he did not feel like making
the long journey, but both he and the Northrup-King people told me that
they would make every effort to supply as large quantities as they could of
high quality seed.

The specifications I used in visiting with them was that we would not
offer to supply any seed for export to your country or to the Danube Valley
countries that was not of as high quality as corn we are selling domestically
in every respect. It will of course be from the same plants as the corn we are
selling American farmers. But it will be only in our choice of kernel sizes.

American farmers found that the profit of planting hybrid corn as
compared with open pollinated corn was so great that they of course plant
nothing but hybrid seed corn, and they try to buy the best in hybrid seed
corn without regard to price. They not only want the best varieties there
are, but they want the popular kernel sizes of those varieties. They want
flat kernels instead of round, for instance, even though there is no difference whatever in the productiveness nor yielding ability between the flat
and round kernels of the same variety. . . .

The average domestic price of the extra corns and the early corns in the
United States is approximately twelve dollars a bushel—and for the medium
maturity strains about eleven dollars a bushel. As I pointed out previously
this is the price of the best hybrids available. Pioneer, DeKalb, and Northrup-King are representative of these very top hybrids. Small producing
companies who do not have the top research, who do not do the top job of
detasseling, who do not do the top job of making their corn germinate
strongly and grow vigorously will be offering hybrids four or five dollars a
bushel less. It has always been so since the start of the industry.

The best proof that complete superiority produces bigger yields for the
customer is the very simple fact that DeKalb and Pioneer and Northrup-
King—that just these three companies alone—will market something like
45 percent of all the seed corn planted in the United States this year and
probably 60% of the total corn marketed in the extremely early and early
corn maturing areas, that is, the great northern Corn Belt.

I visited at length with the Pioneer group, with the DeKalb group,
and with the Northrup-King company, and we discussed the matter of what
price would be entirely fair and equitable on the corn for export to your

country and the Danube Valley countries if you would prove to be interested. With a large purchase and with no serious sales expense on our part you would of course be entitled to a wholesale price. Inasmuch as we say frankly that we will not sell any corn for export that we have a market for domestically to the farmers of the United States, you are buying only surpluses which we have available.

Our sales price domestically contemplates practically no freight cost to the companies involved because the corn is grown in the same areas as the farmers who use it. After a lengthy discussion we decided to offer to sell you whatever corns we could supply at a price of six dollars per bushel for the early and extra early varieties and five dollars per bushel for the medium maturing varieties, a bushel to be fifty-six pounds of corn net above the weight of the bag, the corn to be priced F.O.B. New Orleans and to be paid for in dollars. This is approximately half of the domestic price for the same varieties. We could not for this price give you your choice of kernel sizes—that must be our choice. . . .

It will all be treated with a fungicide exactly as we treat the corn that is delivered to the American farmers, and it will be graded into rather accurate kernel sizes, approximately the same as we deliver to the American farmers. Genetically, of course, it will be exactly as good as the best hybrid seed corn being planted in the United States. It will likely be graded into Large Round, Medium Round, Small Round, and Short Medium Flat and Small Flat. All of those kernel sizes are planted in the United States and planted with great accuracy, and I have samples showing the several kernel sizes we have in mind, which I will submit with this letter. . . .

We want to charge you enough so that we can afford to furnish you only the finest seed and let us have a reasonable profit—beyond that we do not wish to go, because beyond that we would not be deserving of your continued patronage.

Because of the thoughts expressed in this letter, I would appreciate highly your discussing the whole matter thoroughly with your associates at your very earliest convenience. I feel certain that these truly superior hybrids will outyield open pollinated corns by a margin of 25 or 30 percent. The increase will be even larger than that if dry weather handicaps open pollinated corn but not sufficiently enough to prevent it from producing some ears. American hybrids have terrific roots and will go down further after moisture than open pollinated varieties. It is not unusual for hybrids to outyield open pollinated corns in Nebraska and the Dakotas and in eastern Colorado by above 50 percent, whereas in the humid areas of the central Corn Belt a twenty-five percent increase would be more likely.

If you and your associates believe you would look with favor upon the purchase of substantial quantities of these very superior American hybrids as outlined in this letter, then I think it is imperative that I get down into the areas where the seed would be planted at the earliest moment and discuss with your geneticists and actual farm operators the areas in which the extra early varieties and the early varieties and the medium maturity varieties would be best adapted.

Furthermore, if you do not wish to purchase all of the seed that is available for use within the USSR itself, I would like to get over to Hungary, Rumania, and Czechoslovakia and make the same kind of an offer to the peoples of those three countries.

Frankly, I would like to supply at least nominal amounts to those three countries in any case although you were the person who honored me by inviting me to come to Russia and I would feel of course that you and your country have first call against any benefits that I could bring to the whole area.

One other thing I would like to mention. There is practically no open pollinated corn left in the United States, and so we can locate little or no new breeding material from open pollinated strains in our country. In the development of hybrid corns we find it beneficial to use some new and wholly unrelated strains from time to time. I would like, therefore, with your permission, to carry back with me to the United States a few individual ears of your open pollinated corns from fields which I may find that interest me for one reason or another. Maybe it will be because of early maturity, maybe because of some ear type or kernel type, maybe because of a particularly heavy stalk. I cannot anticipate what it might be, but I would like your permission to carry back a few individual ears.

I have written this letter at very great length because of the difficulties of remembering to be sure to cover all of the subject matter and because I thought it would be convenient for you and your associates to have the letter translated into your language for study. I have no doubt at all that much of it will be covered by visiting through an interpreter, but I thought this would give you my thoughts in written form and might be more convenient to you.

I wish to point out that an early decision on whether or not you wish to purchase American hybrid seed corn is of the utmost importance if the amounts that are to be made available to you are to be substantial and that the earlier the decision is reached, the more substantial they can be.

Very sincerely yours,
Roswell Garst

November 1, 1955
To Mr. Vladimir Matskevich,

Dear Minister Matskevich,

When I left you the last time at Yalta I could not make a definite time statement as to when I would get back and, therefore, I could not make a definite time statement as to when I hoped your delegation would be able to come to Coon Rapids. I did return home yesterday and I am, of course, anxious to have them arrive at the earliest possible convenience because I will be able to show them everything more completely before winter sets in.

As per our last conversation, I understand that this group will be authorized to definitely contract for the seed corn according to the letter which I wrote to you September 17 and delivered to you when I came to Moscow. The amounts of total corn by maturities as you gave them to me in Yalta were fifteen hundred or two thousand tons of medium maturity corn, fifteen hundred or two thousand tons of early maturity corn, one thousand tons of extra early maturity corn for a total of a minimum of forty-five hundred tons to five thousand tons and with a possibility that you might want to increase the order but that the above amounts would be minimum.

I understand, of course, and have not neglected giving consideration to the fact that you wish also to purchase inbreds and single crosses for the production of hybrid seed corn via American methods within the Soviet Union. I have not yet had sufficient time to discuss this matter with the geneticists of the several companies but will have done so before your delegation arrives, and I am sure we can be of some assistance in this matter and I am hopeful that we can be of large assistance.

I also know that you wish to buy artificial drying equipment, grading equipment, equipment for treating with fungicides, and all other equipment necessary for the fine production of hybrid seed corn within your nation.

I also know that you are interested in a detailed study of our cattle feeding that I have told you about as we have done it here at Coon Rapids with the use of urea for a substantial part of the protein requirement. For all of these purposes, I understand that you were going to send to the United States a geneticist, an engineer, a cattle feeding expert, and an interpreter. On my part, I want it understood that I agree to be of every possible assistance to all of these men in their particular fields.

In truth, I want you to know that I will consider it a very great privilege to be helpful. Never have I been more cordially received than I was in Russia, and for your information I might add, never have I been more cordially received than I was in Rumania and Hungary—and I am going to consider it a real privilege to be helpful in return. You will, I am sure, be interested in knowing that I invited the Rumanian government to send over a group of three of their citizens to confirm their order for one hundred tons of hybrid seed corn and to purchase several sets—complete sets—of American farm machinery for demonstration purposes in Rumania next year.

And I invited the Hungarian government to send over a delegation of three of their citizens to confirm their purchase of the hybrid corn they wish to purchase and some farm machinery and some hybrid seed corn processing equipment.

When I was in Bucharest and in Budapest, I did in each case call upon the American legation and tell them that I was inviting several citizens from the respective countries to come to the United States and specifically, to Coon Rapids and that they would be asking for visas for this purpose. As you know, it has been unusual in the past ten years for Rumania or Hungary to issue visas to Americans or for America to issue visas to citizens of those countries. The result was that the American legation in Bucharest and Budapest were not in a position to promise that the visas for citizens of those countries would be immediately approved in spite of the fact that they had been very willing to issue visas to both Dr. Schutz and myself without hesitancy and in spite of the fact that they had received us very cordially.

This disturbed both Dr. Schutz and myself—disturbed me enough, in fact, so that instead of going to Czechoslovakia, I decided to come directly to the United States and take the matter up with the State Department. Dr. Schutz went directly to Geneva and was able to visit with the Secretary of State, [John Foster] Dulles, about this whole matter and strongly urged the making of passports, to all of these countries, very general—and the issuance of visas much more general than it had been—urged, in fact, that the greater the traffic between the several countries, the better.

Both Dr. Schutz and I believe that it was sensible to trade information freely on the peaceful uses of the atom as was done at Geneva during the past summer and we both pointed out strongly that if it is wise to trade information on the peaceful use of the atom, it is equally wise to trade information on how to produce more food with less labor and that the best way to trade information was by an interchange of people on specific subjects. I am sure that Dr. Schutz's conversation with Mr. Dulles and my conversation with the State Department in Washington contributed to the decision Mr. Dulles announced in Geneva yesterday, and I am, of course, proud and pleased to have made this contribution. I think it is just as important to the United States as it is to the USSR and that it works to the mutual advantage of both of our countries.

I promised both of you, you and Mr. Krushchev, that upon my return home, I would forward any suggestions that I might have for your consideration, and I now wish to suggest a thought that I did not develop while in the USSR, that I did develop in Rumania.

As I pointed out to you and Mr. Khrushchev, it is my opinion that the corn growing in the USSR will not expand as rapidly as you would both like without mechanization and rather complete mechanization. I also told you both that I thought complete mechanization depended upon having rows one full meter apart and that a very great deal of research in the United States had proved that maximum yields are possible with rows one meter apart. I have written for basic research circulars covering this point and will have it confirmed by such leading American agronomists as Dr. Lambert of the University of Nebraska, who headed the American delegation that visited your country this summer. As you know, practically every bit of corn grown in the United States is grown in rows forty inches wide, which is almost exactly one meter wide. . . .

I told Mr. Khrushchev specifically that I would be happy enough to have him send over a couple of good bright young citizens of the USSR the fifteenth of April and have them stay until the fifteenth of November and show them how one man can produce one hundred hectares of corn and that they could then go back and demonstrate this knowledge in your country. Mr. Khrushchev's answer was that you had quite a lot of territory for two men to educate and that he would like to send over a dozen men, which I thought was a good idea.

I still will, of course, be glad to welcome the two men I suggested and probably could arrange for a dozen to find similarly good places, but the more I got to thinking it over, the more I got to thinking that the suggestion I made to the Rumanian government might be even better—and that it might pay the USSR itself to buy at least a few sets of complete American farm machinery for demonstration within your own country and for study within your own country because you might well find out that the design of the American farm machinery, while it reaches near perfection here, might have to be changed somewhat for maximum use within your own country.

Furthermore, I am absolutely certain that you should give consideration to the most modern production mechanically of potatoes. On our trip through the Ukraine we saw rather large fields of potatoes being dug by hand and picked up by hand with a labor requirement per unit of potatoes very, very much higher than the labor requirment in the United States, and I believe that you have as much to gain from the complete mechanization of your potato culture as you have from the complete mechanization of your corn culture.

We do not produce many potatoes in Iowa, but I have a brother living in California who is very able, and in California they have perhaps the most highly mechanized potato culture in the world, and it will be no trouble at all for me to send a part of your delegation out there for study of how to most completely mechanize potatoes. This brother, Jonathan Garst, is the one who I told you knew Sir William Ogg[1] so well—knows more about fertilizer than anyone I have ever known—is completely sure that the best way to improve relationships between the USA and the USSR is to trade information on how to produce food more effectively and with less labor—and is a person whom your delegation should know.

You, yourself, having visited the United States under the glare of the utmost publicity because you were the first delegation to visit the United States, must know how impossible it is to get much done in the way of actual accomplishment when surrounded by correspondents, photographers, and just simply curious friendly people, and you, yourself, must know how fatiguing such a trip becomes.

For that reason, I think little or no publicity should be put out concerning the men you are sending on this mission until at least toward the close of their visit to the United States. I will see, of course, that they meet every agronomist whom they need to meet, all of the machinery people that they need to meet, that they get every possible opportunity to learn the maximum amount and to purchase not only the hybrid corn that is best adapted to your use but the machinery that is best adapted to your use and, in every way, be of full help to them. But I do hope that any large amount of publicity about their mission be held up until they have finished all of their business and all of their acquisition of knowledge, after which time I think the maximum amount of publicity would be excellent. . . .

I trust you will discuss this letter with Mr. Khrushchev, who was, himself, interested in any suggestions I cared to make after my trip. I urge you both to give serious consideration to the purchase of several complete sets of agricultural machinery for row crops and I specifically suggest not only complete corn machinery but also some complete potato harvesting machinery.

And finally, in closing, I want to repeat that it will be a real privilege

---

1. Director of the Rothamsted Experimental Station, Hertfordshire, England.

for me to be just as helpful as possible in every way to your delegation. Thanks once more for the great cordiality extended to Dr. Schutz and myself while we were in the USSR.

Very sincerely yours,
Roswell Garst

December 12, 1955

To Mr. I. E. Emelianov, and
Members of the Russian Delegation

Gentlemen:

I am writing this letter at some length because you came at my invitation and because I believe you will want this advice written down so there will be no mistake about it.

I expect that tomorrow you will arrive here by before noon and that we will have time tomorrow to conclude the actual purchase of the seed corn itself because it is necessary that we get this contract concluded immediately in order to get the sizing, tags, and supplies arranged and acquired and the shipping arranged for.

But while I am sure that this matter can be taken care of tomorrow, I am almost completely sure that there is not sufficient time remaining under your visa time limit to conclude all of the things which you wish to have done.

I presume, therefore, that a good deal of other matters will, of necessity, be handled by Amtorg[1] after your departure from the United States.

Export licenses will have to be acquired, not only for the seed corn, but for each of every item that is acquired in the United States. The one man that has represented us with the State Department has been Mr. Philip Maguire of the firm of Becker & Maguire, Attorneys of Washington, D.C. . . .

But the time has been simply too short to conclude all of the things that need concluding. I simply think it is necessary for Amtorg to represent the USSR beyond your visa date and for some one man to represent all of the American suppliers beyond your visa data, and I think that one man to represent all of the American suppliers simply has to be Mr. Phil Maguire unless great confusion is to be endured.

---

1. The Soviet trading agency, located in New York City.

The State Department has given Mr. Maguire pretty strong assurance that anything purchased as a result of your trip, which was at my invitation, would be granted an export license. I think it is only natural to . assume that they do not want to have seventeen different people coming over to the State Department asking for export licenses for twenty-seven different articles and claiming that the sale was a result of my trip.

I therefore suggest to you and, with your permission, will suggest to all of the suppliers that the American suppliers have Mr. Philip Maguire, or the firm of Becker & Maguire, represent them and that you have Amtorg represent you, so we get it into a sensible procedure.

Now when it comes to the plants you are building, I think it will be essential for you to have one outfit plan the whole thing by getting an engineering firm to make the layout of the buildings, and then having everyone connected with it take care of their own particular specialty: for instance, the Finco people to design the receiving equipment, the husking beds, the sorting tables, and the conveying equipment as well as the corn shelling operation. The Campbell Heating Company should design the fans and heating units necessary for the drying of the corn, along with a sketch of the dryer type to be used. The Superior Separator Company should plan all cleaning and sizing equipment and should outline a complete sizing, treating, and bagging plant.

Because your whole delegation will be returning to Moscow in about a week, I think you might be well advised to take this letter, plus all of the estimates on all of the machinery, plus all of the estimates that these three companies will jointly make, back home with you and discuss them at length but in a very short time, with Mr. Matskevich, and then authorize Amtorg to proceed—and I am completely certain that the American suppliers can work through Mr. Maguire so far as personnel and export licenses are concerned, and that there ought to be no difficulty. . . .

I do wish you to know that supplies of everything are short in the United States in relation to the demand. Steel is short, labor is short, business activity is at a new high. I think it will be difficult to get delivery from the Superior Separator Company, or the Campbell Heating Company, or the Finco people, or the Oliver[2] people in less than 120 days, or maybe even 150 days. It takes a month by boat to get it transferred over. You only have nine months before harvest. So there can be no delay at all on getting the thing under way once you have had time to take the matter up with Mr. Matskevich.

2. Apparently the Oliver Construction Company, Oconomowoc, Wisconsin, commercial building and warehouse contractors.

You simply must impress Amtorg with the fact that they simply cannot quarrel over small details. The American manufacturers will be going out of their way in supplying this material out of already busy schedules. They will be doing it, in my opinion, only because of the hopes of future and larger orders. They will not be patient with Amtorg if there is any tendency to delay and quarrel about small details, although they will be most cooperative if they get good decisions and rapid decisions and firm decisions. I think it is of special importance that Amtorg understands that this material must be authorized at the earliest possible moment because everything has to be specially built to fit a particular plant design.

The same thing is true with the farm machinery being purchased from the International Harvester Company and others. For instance, when I was in the International Harvester Company, I found out that Amtorg had ordered very small amounts of equipment—a couple of tractors, a couple of plows—and that Amtorg had forwarded to the International Harvester Company a contract form that provided for adjustment of any dispute in Europe and had a lot of qualifying clauses and that the International Harvester Company sent the order back and simply said it was not worthwhile to them to make special details for two tractors. . . .

It is this type of thing that simply cannot be allowed to happen in the purchasing of equipment for the plants which you expect to build. I am sure that Mr. Maguire will draw up contracts for the local suppliers that are completely fair and that he can convince the Amtorg people that they are completely fair, but certainly the Amtorg people must be instructed by your government not to quarrel about the small details, or you won't be able to get the plants built in two years' time.

The American suppliers have so much business, as I pointed out once before, that they simply will not do business with anyone quarreling about small details and insisting upon arbitration in Europe—they will just give up the whole deal rather than be bothered. . . .

This letter has become long enough but I cannot close it without again expressing the thought that it has been a pleasure to have you all come to the United States. It has been a pleasure for all of us here to be of assistance to you. It is a pleasure for me to write this letter of advice.

I am strongly of the opinion that exchanges of this kind are the best way of increasing the chances of permanent peace in the world. I hope and expect that as a result of my trip to the USSR last fall and your trip to the United States now, greater trade will develop between our two countries.

As I told Mr. Khrushchev and Mr. Matskevich and Mr. Mikoyan during the pleasant afternoon I spent with them, I believe there would be a real advantage to both countries in opening up a much wider trade. Lard is six times as high priced as bread in the USSR. In the U.S. it is the same price as bread. So it seems to me that you could use lard and perhaps meat and dairy products and maybe even eggs, which we have in too long supply. . . .

I am proud of what I believe to be a rather substantial contribution between our two countries. I am hopeful that this small contribution may open wider opportunities. I shall never forget Mr. Emelianov, the thoughts that you expressed at the meeting of our Commercial Club here at Coon Rapids, when you said that you thought of the seed that your government was purchasing as being not only the seeds of maize, but perhaps the seeds of friendship. I think I never heard anything more appropriately said than that.

I also hope that they will prove to be seeds of friendship.

May I wish you all a fine journey home and assure you that you go with the best wishes of us all.

Very sincerely yours,
Roswell Garst

June 28, 1956

To Congressman Harold D. Cooley,
Representative from North Carolina, and
Congressman Victor L. Anfuso,
Representative from New York

Gentlemen:
. . . As you both know for the last several years the Commodity Credit Corporation has been debarred from selling agricultural surpluses it owned anywhere behind the iron curtain. President Eisenhower asked in his January message that this situation be changed so that the Commodity Credit Corporation could sell behind the iron curtain—and I believe Secretary Dulles recently appeared before the Senate Agricultural Committee urging prompt action on this request.

I personally believe that there would be quite a market for our surpluses behind the iron curtain. For instance, fats are very, very scarce. When I was in the USSR lard was eight times as high per pound as bread, whereas in the United States they were about the same price. Cottonseed oil and soybean oil could be used in terrific quantities not only in Russia but in Rumania, Hungary, Czechoslovakia, Poland, and East Germany. Clothes are very high in price because of a shortage of cotton all over that area. Hides are so high that it takes a man a week to come by a pair of shoes, whereas in the United States one day's work will always earn a pair of shoes. They have some minerals and metals that we could use—they are exporters of cement, which is in very tight supply here—and there are certainly some products that they could supply us in exchange.

And I am quite certain that the better look they have at our progress, the less chance communism in its present and past forms has of continuing. They are so far behind the U.S.A. that I am quite sure they will gradually change from dictatorship toward democracy when they see the advantages that democracy has as a way of life.

Small delegations from the USSR and Rumania and Hungary did come over to inspect the seed corn we shipped them last winter and to purchase machinery and to conclude the transactions which I had with them. Each delegation spent a little more than a month in this area under my guidance.

We showed them supermarkets and of course it pretty nearly drove them crazy. We showed them four-row planting equipment, four-row cultivators, two-row corn pickers—all of the American modern farm machinery. We showed them tenant farmers living in fine, modern homes, hired men living in modern homes on farms, the workmen's houses in town which of course are just as modern as the home of the owner of the business.

We did to a few people from each country, what an American exhibit could do to hundreds of thousands of people in Moscow—stimulated their desire for better living and proved to them that American democracy had brought us the highest living standards in the world. We didn't talk politics at all—you don't need to when you can show them.

I think the United States can do a wonderful job of diplomacy by having an exhibit in Moscow[1] and that any small expenditure made to see that it is a very excellent exhibit is the greatest benefit that could happen. One bomber costs several million dollars and one extra bomber isn't going to scare them very much.

I suspect that a few hundred thousand dollars spent on a really good exhibit in Moscow brings peace closer than several million dollars spent on one extra bomber.

Furthermore, I hope that at this session of Congress you do authorize Commodity Credit to sell American agricultural surpluses behind the iron curtain. I think they would furnish a terrific market for cotton, fats, wheat,

---

1. The International Trade Show would be held in Moscow in the summer of 1957.

corn, and protein. I am extremely critical that the Department of Agriculture itself—Secretary Benson himself—has not appeared before your committee.

However, he has not been a good secretary of agriculture anyhow, in my opinion, so I am disappointed without being surprised.

I congratulate you both on your interest in the possibilities of an American exhibit in Moscow and urge that you use every effort to promote such an exhibit. I think it would be the finest thing we could do.

Very sincerely yours,
Roswell Garst

February 12, 1957

To Mr. Vladimir Matskevich
and Mr. I. E. Emelianov

Dear Mr. Matskevich and Mr. Emelianov:

I have had no contact with either of you since mid-November of 1956 when I cabled that, due to the very disturbed international situation, it would not be advisable to have delegations from the USSR visit the United States at least until the situation improved.

The use of the armed forces of the USSR in Hungary caused very bitter feeling on the part of the American public toward the USSR, and at the height of that resentment in November, December, and early January there would simply have been a too great influence of resentment and too great a chance of incidents against visitors.

That resentment still exists, but is much less pronounced now than it was thirty or sixty days ago. It would have been impossible for us to have sold corn to your government during that whole period because of dangers that it would not be loaded or unloaded at the docks—and certainly the chance of unfavorable publicity about trade would have every likelihood of having been discussed in the American Congress.

I doubt if you can possibly realize how critical a situation was created. Before the Budapest affair there had been a very great increase in the friendly relations; there was a very violent reaction to Budapest and the subsequent events in Hungary, and only now is the situation starting to improve. I hope very much that it will continue to improve and that the situation can get back to where it was before Budapest at the earliest possible moment—and then move from there into a friendlier atmosphere.

My ambition now—as it has been ever since I first met Mr. Matskevich—is to do everything in my power to improve the relationships between our two countries. I think I did exactly right in not doing a thing nor saying a thing until I thought the time had come to start going forward again.

I never did think of terminating the present relations I have had with you men and your associates. I just thought it was wise not to do a thing until the situation improved to a point where I thought they might go forward again.

It is, of course, too late for you to purchase any seed corn for planting this spring. Shipping is extremely tight—and slow. The corn that we sent to the USSR last year was shipped before this date and it arrived, even then, almost too late. So, I think it is impractical to consider any seed corn business this year.

Anyhow, you did carry over toward half of the corn we sent you last year, so you can continue planting Pioneer this year, using the stock that you carried over and have on hand.

We are materially expanding our production of hybrid grain sorghum seed, and I do want you to have the opportunity to check and study hybrid grain sorghums across the USSR in 1957. With that in mind, Mr. [Leo] Schneider[1] and I have prepared forty five-pound samples of hybrid grain sorghum seed, and I stopped at the USSR Embassy in Washington and visited with Mr. [Sergey Romanovich] Striganov,[2] and he has assured me that he can get this limited amount sent to Moscow by passenger boat and by passenger train so that you should receive it early enough for wide distribution.

I am also forwarding to you a few samples of the granular fertilizer that I told you about when we were down on the Black Sea. Granular fertilizer is just about as easy to make as the old powdered types—and infinitely easier to use because it never cakes and always stays in fine physical condition which makes it easy to spread—and easy to use in mechanical equipment. I thought you should see samples of the product as we use it here in the United States.

I am extremely anxious also to see you get started in the production of broilers. By broilers I mean young chickens that are raised up until they weigh about a kilo and one-third—and then butchered and fried. They are one of the finest meats in the United States and chickens produced for meat in this way are extremely efficient. Big commercial operations are the usual

1. Chief of the germination laboratory at Garst & Thomas.
2. Counselor in the Soviet Embassy in Washington.

kind. The chickens are actually butchered at about ten weeks of age and they will produce a kilo of live weight for about every 2.7 kilos of feed, if the feed is right. . . .

I am, as I always have been, extremely anxious to see better relations between your country and my own. That is uppermost in my mind.

I think better relations are more likely to come about through increased trade between our two countries than any other way. . . .

Very sincerely yours,
Roswell Garst

June 15, 1957

To Mr. V. J. Tereshtenko

Dear Valery:

I have, frankly, been in confusion and hoping that the confusion would end, but it hasn't entirely ended. . . . I wrote a letter to Zaroubin a week ago, studied it a week, and then threw it away.

I think I will write him another letter in a few days, after I have further contemplated the whole situation. I like Matskevich. I like Emelianov. I even have respect for the directness of Khrushchev. I think I know their problems better than they know them themselves.

They want to increase food; they want to raise the standard of living of the Russian people; they all know that these things must be done to prevent a blow-up, and that within their own country.

I think they are unnecessarily worried about an attack on them from the West—but I think they are very worried about it anyhow, and I can understand why. Our Congress and our senators and our secretary of state are just as worried about Russian encroachment into the so-called free world area. The reestablishment of mutual confidence is a slow process, a delicate one, and one of very great importance.

So I do not apologize for being confused, for thinking the thing through very carefully, nor about moving forward with hesitancy.

But let's review where we stand at the present time, both for your thinking and my own.

I think up to now the things we have done have been exactly right. I would not change any of the things I have done in the past. I think probably that that's the best guide on to how to proceed in the future—to just be quietly helpful and courteous and decent, to do business with them just as I would do it with my domestic customers, to be helpful in showing them how to raise the biggest corn crop possible, and to show them how to use that corn with greatest efficiency. . . .

Sincerely,
Roswell Garst

January 2, 1958

To Mr. Chas. Thomas
Mr. M. E. Galloway
Mr. James Wallace
Mr. Nels Urban
Mr. Wayne Skidmore[1]

Gentlemen:

After visiting with you the other day about the exporting of Pioneer hybrid seed corn, and particularly about exporting Pioneer hybrid seed corn to the Soviet Union, I came home with a very hollow feeling and didn't sleep very well because I couldn't help thinking that maybe a mistake was being made.

So, the day before yesterday, I asked Wesley Thomas[2] to get me up the figures so I could look at the whole picture, and I think it's worthwhile to re-present the whole matter via this letter. The figures shown below cover all the sales actually made behind the iron curtain in the last three years. They are as follows:

PIONEER OF DES MOINES

1956 Russia
15092 Bu. @ $5.00 $ 75,460.00
64088 Bu. @ $6.00 384,528.00
120 Bu. @ $120.00 15,000.00
$474,988.00

Poland
25700 Bu. @ $6.00 $154,200.00

GARST AND THOMAS

Hungary
2300 Bu. @ $6.00 $ 13,800.00
1500 Bu. @ $5.00 7,500.00

Rumania
29527.5 Bu. @ $5.00 $147,637.50
9842.5 Bu. @ $6.00 5,905.50

---

1. Thomas and Galloway were Garst and Thomas executives; Wallace, Urban, and Skidmore were Pioneer Des Moines executives.
2. Business manager for Garst and Thomas.

				Russia			
				25960 Bu.	@	$6.00	$150,360.00
				54301 Bu.	@	$5.00	271,505.00
1957	Rumania			Rumania			
	1920 Bu.	@	$5.00 $ 9,600.00	80896 Bu.	@	$6.00	$404,480.00
	3948 Bu.	@	$6.00 23,688.00	179 Sorg.	@	$12.32	2,205.28
	100 Bu.	@	$134.00 13,400.00				$406,685.28
			$ 46,785.00				
1958	Rumania			Rumania			
	9276 Bu.	@	$5.00 $ 46,380.00	41952 Bu.	@	$ 5.00	$209,760.00
	550 Bu.	@	$100.00 55,000.00	30 T. Sorg.	@	14¢LB	9,259.32
			$101,380.00				$219,019.32

The grand total has been something above 365,000 bushels of commercial corn, 770 bushels of parent corn, 1,360 bushels of sorghum. The grand total in dollars is $707,356 for Pioneer of Des Moines, $1,222,412 for Garst & Thomas, for a total of almost exactly $2,000,000. . . .

In short, the commercial aspects of what we have done are, I believe, not subject to much question. The foreign business has been extremely profitable, in my opinion—and it can continue to be so, in my opinion.

That gets it down, then—if you agree with not too great violence to the foregoing—to a moral or a patriotic or a sentimental reaction. And I recognize that there is a difference between your feelings in these matters and my own.

So I want you to know how I feel about the matter as clearly as I can spell it out.

In the spring of 1955 the *Des Moines Register* had invited the Soviet Union to send over a delegation to find out how to run an Iowa Corn Belt. The arrangements had been largely made for the exchange of farm delegations. I was in Washington in April of 1955, and I got to thinking how much fun it would be to have some plots of Pioneer hybrid seed corn in Russia when the American farmers arrived there.

So I decided to go over to the Russian Legation and suggest the matter to them. I was advised by my friends not to go to the Russian Legation because [the] CIA had long-distance photostatic equipment they used to take pictures of everybody that went into the legation. I suggested calling the Legation and was advised that all the telephone lines that went into the Legation were tapped. I called Dick Wilson[3] of the *Des Moines*

3. Chief of the Washington bureau.

Register and offered to furnish seed if the *Des Moines Register* would attend to the arrangements, and even Dick Wilson shied away from that. It is perfectly impossible to exaggerate the fear that people had in Washington of knowing anybody who knew anybody who was a Communist. McCarthyism had its pinnacle is the only description I think appropriate.

The exchange of farm delegations was the opening wedge. The American farmers were well received in the Soviet Union. The farm delegation was well received in the United States.

I had to maneuver to get the Soviet delegation to visit Coon Rapids over the protests of Iowa State College, but they did get here and you know the contacts pretty well since that time.

I simply have to add that as early as March and throughout the spring and summer of 1955, I was corresponding with Senator [Bourke] Hickenlooper, with the Department of Agriculture, and with others and pointing out that American farm production was certainly fifteen years ahead of our population; that iron curtain farm production was certainly fifteen years behind their population; and that finally, with the atomic and hydrogen bombs, that wars were now impossible; that we probably could more effectively use 30 billion dollars worth of armaments and 5 billion dollars worth of food, better than we could use 35 billion dollars worth of armaments.

That has always been my attitude—it's still my attitude and will continue to be my attitude. . . .

Inviting delegations over to the United States to confirm purchases of seed corn started opening the thing up. The trips of David Garst and his wife, [and of] Bill Brown, Mrs. Garst, the Mathyses, Tereshtenko, and myself did much more. I talked constantly about the desirability of greater movements of people back and forth between the two areas—and the improvement has been nothing less than phenomenal.

From no Americans going to Russia in 1954—and a dozen in 1955—some 2,500 went in 1956 and even more in 1957. From no Americans going to Rumania ever until 1955, more than 250 went in 1956 and nearly 1,000 in 1957. Where there was bitterness between the American Legation in Rumania and the Rumanian government in 1955, there was great cordiality when Bill Brown and I were there a year later—and the greatest kind of cordiality with the two when I was there two weeks ago. . . .

I am proud to feel that I have made a very major contribution to the lessening of tensions between the two areas—it has been one of the most satisfactory experiences of my life.

With hydrogen and atomic bombs, wars are now impossible. A major war now would end civilization as we know it. I am completely certain that all of Europe, including the Soviet bloc—having experienced the actual tragedy of cities destroyed, devastation, occupation, the real horrors of war—fear war much worse than the United States. I see no likelihood of war unless——

Unless the two areas continue their respective fears, suspicions, hatreds, and continue building armaments higher and higher until the burden gets so great that they decide it is not worthwhile. I must put one statement in about a word I used—the word *hate*. I know a great many people in the United States who "hate" the Russians, or communism. I never heard anyone behind the iron curtain use the word *hate* in any connection except one—they "hate" war.

Anyhow, that's enough for my general attitude. What do I suggest?

Well, frankly, I would like to continue doing business with the Soviet Union. I think it can be done with profit, I think it can be done with high patriotic or moral goals. I think it is in the best interests of our two companies to have it continue. . . .

Sincerely,
Roswell Garst

January 10, 1958

To Mr. V. J. Tereshtenko
Hotel Metropol
Moscow, USSR

Dear Valery,

It is with deep regret and only after very careful consideration that I write this letter.

But until a further improvement in the cordiality in the relationships between the American government and the government of the Soviet Union takes place, Garst & Thomas do not feel it advisable to supply the seed for planting within the Soviet Union.

A year ago directly after Hungary, the feelings of the American people were extremely bitter toward the government of the Soviet Union. Trade between the two countries at that time would have been severely criticized by a very high percentage of the American people.

Conditions are much improved and I hope and expect they will continue to improve. But at the moment, we do not believe they have improved sufficiently. The accusations that the Soviet government made that the United States was encouraging Turkey to attack Syria last summer were a second severe blow to the improving feelings between the two countries.

The verbal brickbats that have been hurled back and forth between the two governments—and I certainly agree that the hurling has been done by both governments—the verbal brickbats have just created an attitude and an atmosphere in which it is uncomfortable to do business, and not worthwhile.

Although we feel we should not, at this time, proceed in a business way, I will be most happy any time after February 15 to receive a delegation and be just as helpful as possible to them. But we have scheduled meetings in the last half of January and the first half of February clear across the territory which we serve, and so I will not be at liberty until after February 15.

I am sure you, personally, realize how disappointed I am to feel that I must write a letter of this kind. I am not at all sure our judgment is right. I can only assure you that it has been arrived at after very serious consideration. . . .

I was very sure a year ago, after Hungary, that the decision not to sell corn to the Soviet Union at that time was absolutely correct not only from the standpoint of Garst & Thomas—but from the standpoint of the Soviet Union itself. Although the State Department and the Department of Commerce would have issued an export license even at that time, the feelings of the people were so very bitter that trade might well have encouraged antitrade laws.

The very fact that there was no question a year ago, that there is a very serious question as to whether we have made a correct decision at this time or not is proof of the fact that conditions are very materially better.

And so I am hopeful that 1958 will bring a further improvement so that we can, a year hence, reverse our attitude and feel really good about it.

Some progress on arms reduction would be a wonderful help: the easing of tensions between the two countries, the greater interchange of people between the two countries, the cessation of threats being hurled back and forth, cooperation on things that the two countries can work at jointly.

I thought the exchange of new year's greetings initiated by the top Soviet officials and answered in kind by President Eisenhower got 1958 off to a fine start.

I hope, and I believe, that progress will be made. I certainly will contribute everything to that progress to the best of my ability. . . .

Very sincerely yours,
Roswell Garst

JANUARY 14, 1958

CABLEGRAM TO VALERLY J. TERESHTENKO
HOTEL METROPOL
MOSCOW, USSR

REVERSING THE POSITION I TOOK ON JANUARY 10 GARST & THOMAS ARE IN A POSITION TO SUPPLY PIONEER HYBRID CORN AS FOLLOWS STOP 75 TONS OF PIONEER 371 STOP 225 TONS OF PIONEER 349 STOP 335 TONS OF PIONEER 352 STOP 300 TONS OF PIONEER 347 STOP 35 TONS OF PIONEER 345 STOP 180 TONS OF PIONEER 344 STOP 75 TONS OF PIONEER 325 STOP 375 TONS OF PIONEER 329 STOP 475 TONS OF PIONEER 339 STOP 375 TONS OF PIONEER 335 STOP 325 TONS OF PIONEER 318 STOP VARIETIES ARE LISTED FROM EXTRA EARLY TO MEDIUM IN ORDER OF MATURITY STOP.

WE ALSO HAVE AVAILABLE 750 TONS OF PIONEER 300 STOP 500 TONS OF PIONEER 332 STOP 250 TONS OF PIONEER 312A STOP 250 TONS OF PIONEER 302. THESE LAST FOUR VARIETIES WOULD ONLY MATURE SOUND GRAIN IN GEORGIA OR UZBEKISTAN OR OTHER TERRITORIES OF THAT NATURE BUT THEY WOULD MAKE VERY GREAT QUANTITIES OF SILAGE IN THE ODESSA-KRASNODAR AREAS OR SOUTH OF KISHINEV. . . .

OUR SUPPLIES OF GRAIN SORGHUM SEED ARE LIMITED AND GERMINATION WAS HURT BY THE FREEZE. WE CAN AND WILL SUPPLY UP TO ONE HUNDRED TONS ON THE BASIS OF ELEVEN CENTS PER POUND WITH THE GERMINATION TO BE ABOVE 70 PERCENT.

WE COULD HAVE BOTH THE CORN AND THE GRAIN SORGHUM AT THE PORT BY FEBRUARY 15 IF THE ORDER IS PLACED BY CABLE ALMOST IMMEDIATELY. KINDLY ACKNOWLEDGE YOUR RECEIPT OF THIS CABLE. I TRIED TO PHONE YOU LAST NIGHT WITHOUT SUCCESS.

ROSWELL GARST

March 24, 1958

To Mr. Valery Tereshtenko
Hotel Metropol
Moscow, USSR

Dear Valery,

We have had a terribly busy season so I have not had too much time up until now for philosophic thought. But today I received a report from the National Planning Association on Soviet agriculture both past and future which I read with great interest, and it made me think a little philosophically.

This report pointed out what Mr. Khrushchev has said repeatedly in the last several years—that a present major goal of the Soviet government is to improve the diet of the Soviet people to where it will be as good as the diet of the American people and to do that as rapidly as possible. I wish to quote a paragraph from the report:

"The drive of the Soviet Union for better diets for the people is wholly commendable. If it is carried out vigorously with a serious effort to reach the Khrushchev goal, the Russians will have to divert resources away from heavy industry and military preparation toward agriculture. The program will also require more production of consumer goods to provide incentives for farm people.

"This is a 'butter instead of guns' policy and, *if really carried out,* should be applauded everywhere."

Now, as you know, I have always and sincerely believed that this was the goal of Mr. Khrushchev, of Mr. Emelianov, of Mr. Matskevich, of everyone that I came in contact with within the Soviet Union—I have always believed that their greatest ambition was to produce a materially better diet and do it with a smaller percentage of their population than has ever been true in past years or past centuries. While a great many people in the United States have feared that the real objective of the Soviet Union has been to conquer the world, I have never believed this—I have believed that what they really wanted was the best possible standard of living. I have always believed that the warlike gestures and the great armaments of the Soviet Union were based upon fear that someone else would try to conquer them as the Germans have tried to do twice in my lifetime.

Now, with atomic and hydrogen bombs, war is simply unthinkable. Upon that, I believe everyone is agreed. I think war as an instrument of international affairs is completely obsolete. That's why I think disarmament is sure to get started and progressively increase.

But let's just look at what is necessary if Soviet agriculture is to approach the United States. In the United States, we are currently consuming more than 220 pounds of meat per person—in the Soviet Union, just above 70 pounds. We probably consume something like equal amounts of butter per person, but the United States produces and consumes more margarine than they do butter so fat consumption is probably twice as high in the United States as it is in the Soviet Union. And certainly, milk and eggs consumption in the United States is materially higher than in the Soviet Union.

In short, if Mr. Khrushchev's stated goals are even to be closely approached in having as good a diet for the Soviet Union as the United States currently enjoys and of accomplishing that in the next few years—if these goals are even going to be closely approached, the Soviet Union will have to make a terrific drive in all sorts of new expansions in agricultural production.

And this simply cannot be accomplished without extraordinarily large capital investments. It's going to take huge investments in nitrogen production facilities. It's going to take huge investments in farm machinery. It's going to take huge investments on irrigation, in fencing, in feed manufacturing plants, in chemical plants for insecticides and herbicides, in laboratories for the production of antibiotics, and in farm machinery. Now, Valery, I think it's possible to make extraordinarily great progress—much the most progress that this world has ever seen—but it's going to take terrific doing. . . .

I am just as anxious to be helpful as I ever was—probably even more anxious than I have ever been before. This is because I see the burden of armaments finally being recognized everywhere in the world and because I see impending a shift from the armaments race to a higher standard of living on a global scale.

Because many people in the United States, if not most of the people in the United States, still believe that the Soviet Union's main objective is the conquest of the world, the situation remains difficult. A showing of real good faith and real desire on the part of the Soviet Union at any meeting that may be held this summer would be the biggest step forward that would be possible.

I know that the United States and her allies are also entitled to try to make such a conference useful in the lowering of tensions and in moving toward a really substantial reduction in the armaments burden.

Until such a time as headlines on both sides become more peaceful than warlike, little progress can be made. But the contribution that American agriculture can make to the Soviet agriculture will be very phenomenal when conditions get right.

When that time comes, I cannot predict. But in the meantime, I have such utter confidence that the time will arrive that I will be very glad to give advice in a quiet fashion even before the time arrives. For instance, at this time, I would not like to entertain a sizable delegation from the Soviet Union with the arrival of that delegation highly publicized and with lots of headlines about it. On the other hand, I would be delighted to have Mr. Emelianov and perhaps one or two others arrive with a minimum of publicity so that I could give him the maximum of advice with a minimum of publicity. . . .

I hope you will submit to Mr. Matskevich and Mr. Emelianov and through them, at least, to Mr. Khrushchev my broad thoughts on the problem that faces them and my kindest personal regards, and I hope you will do this at your very earliest opportunity and let me hear from you as to their reaction. . . .

Sincerely yours,
Roswell Garst

May 7, 1958

From Valery Tereshtenko
Moscow

Dear Mr Garst,

. . . There is another point, however, regarding which I would like to be more sure than I am at the moment (unless your coming letter, with which this will cross, will endorse Mr. Emelianov's plan as I pictured it in my last letter of April 9th), namely regarding the length of the stay and the purpose of the delegation. They made here a "sour face" when they learned that your invitation is for two weeks only, to be continued for another two weeks by invitation of Mr. Finley.

As to the latter, I feel that there may be some confusion in the whole plan. It is true that Mr. Emelianov wishes to see Mr. Finley too. The other two "men from the soil", however, may hardly be of any interest to Mr. Finley: they are "field workers" specialized in agricultural production; they do not know anything about the feed-mills; at best, they may be interested in a short visit to Aurora to see some of Mr. Finley's agricultural machines. In the meantime Mr. Finley probably expects to meet some Soviet specialists with whom he could speak on feed-mills. The latter matter is not on the program of this particular delegation (the American-Soviet Agreement provides specifically for a delegation of feed-mills specialists).

As you know, the position of Mr. Emelianov is unique. Of the three men under consideration he is the only one who may talk in terms of the American-Soviet trade relations, discuss the question of further purchases of your hybrids (although in May, of course, it will be hard to speak on this), and meet again with some of the agricultural stations' corn breeders. Neither he nor Matskevich would like to make, however, of the visit of the other two men a kind of a two-week "sightseeing" trip as they say.

At the time of your conference with Mr. Khrushchev, pointed out Mr. Matskevich, you promised to give to the Russians a certain *training* in your methods of work and provide them with a chance to get some *practical experience* in them. Next year you promised the same to Mr. Matskevich when you were his guest, says he. And precisely for this purpose they wish now the above two field-men to go to the U.S.A., rather than for any "sightseeing" or talks.

I explained it to you in my last letter how they visualize the procedure of getting "on tractor". They don't want to be a burden to you; they wish you to assign them to work on one of [the] farms, or they may rent 2 rooms (if Marina comes too) in a Motel in Coon Rapids, and quietly stay there— leaving and coming for each new phase of the field work, *working* together with your people. In other words Matskevich wants them to undergo a practical training along the same lines as this has been arranged for two Poles, according to a newspaper article of February 1, 1958 which I read the other day, by the Pioneer Co. (of course, the Russians do not need to attend also an agricultural school, as do the Poles). It is too late to discuss this matter by mail, but perhaps the visa could be extended and any necessary arrangements could still be made.

You know me, Mr. Garst, as a very careful man in such matters. Before I "cut off" anything, I like to measure and to deliberate long, sometimes even too long. Yet this time I am ready to plead strongly for Mr. Matskevich's request. I feel that the fulfillment of his request is something which is absolutely necessary as the first practical step towards the implementation of the whole philosophy you formulated in your letter of March 24. Also, such an action will bridge the gap in your relations with the U.S.S.R. which seemed to have appeared after the unfortunate events in Hungary. . . . [1]

With best regards to Mrs. Garst and yourself

Sincerely yours,
Valery Tereshtenko

---

1. Alexander Gitalov, Vassily Shuydko, and interpreter Marina Rytova spent the summer of 1958 in Coon Rapids.

July 3, 1958

To Mr. Emelianov and other members of the delegation
Coon Rapids, Iowa

As you know, I have now spent nearly full time with your delegation since you arrived on June 18, and I do expect to spend all day Saturday and until you leave Sunday afternoon. I have arranged for you to see detasseling being done in Missouri Monday, July 7, and I have arranged for you to drive toward Lincoln, Nebraska, the afternoon of July 7 so that you can arrive at Lincoln by 10 o'clock in the morning to visit with Dean Lambert. I just talked with Dean Lambert and he expects you in his office at 10 A.M., Tuesday, July 8.

Mr. John Strohm[1] is anxious to visit with you, and I think that I can contact the University of Wisconsin and find out that they would like to have you either July 11, 12, and 13 or maybe 14, 15, and 16. In any case, between the University of Nebraska and the University of Wisconsin and and John Strohm, you will be fully occupied until July 17 when the arrangements have been made for you to get together with Mr. Finley, who expects to not only show you his establishment but to take you to the University of Illinois, Michigan State University, and to deliver your delegation to Washington, D.C., by July 22.

I will have to terminate my visit with you the afternoon of July 6. I have all of my sales supervisors invited in for July 9 from Missouri, Kansas, Oklahoma, Colorado, Nebraska, and Iowa, and I simply must have July 8 and July 7 for the preparation and organization of the agenda of this meeting.

So I wish now to summarize my total recommendations. That way we can spend Saturday and Sunday morning discussing these recommendations in detail.

First, let's take the machinery. I would purchase all of the machinery we visited about the other evening with Mr. Grettenberg and with Steve and Dave and John Chrystal. It is my mature judgment that each one of those sets of farm machinery is very ample in size to farm at least eight hundred hectares—two thousand acres. It may interest you to know that these will be the best equipped farms that I would know of because all of the machinery will be new and modern.

I hope you will have seen the farmhand loaders for the building of haystacks and straw stacks—and for the tearing down of the haystacks and straw stacks for the loading of it into hay racks for hauling. They are a great tool and I think I would buy from Superior Separator Company one or two

---

1. Agricultural journalist who was host to the Soviet agricultural delegation in 1955.

of these machines for each set of farm machinery. I also think it might pay you to buy at least a few feed wagons from Superior. They make the best feed wagons available.

The Olin Mathieson Chemical Company are unwilling to sell fertilizer to the Soviet Union. They are however, willing to sell it to me or to Mr. Grettenberg—pack it in plain bags without their own insignia—but mark it with the words "Ammonium phosphate 16-20-0" or, in the case of the starter fertilizer, they are willing to mark it "Starter fertilizer, nitrogen content 8%, phosphorus content 32% potash content 16%" etc.

Because I am going to be so busy and because Mr. Grettenberg is willing and will have time, we have agreed that he will get the export license for not only the machinery which you wish to purchase from him but also for the fertilizers that you wish to purchase.

Mr. [Bowen] Campbell[2] is quite willing to get his own export license and take care of your requirements as per his letter. Mr. Finley, of course, is quite willing and anxious to take care of his own export licenses and sell directly to you. So will the Superior Separator Company be willing to do the same.

I hope that by the time you get to Washington, Mr. Grettenberg will have been able to fulfill all of your requests for circulars, information, etc., that Mr. Finley will have been able to do the same—and that you can get all of your final contracts negotiated.

Fortunately, Mr. Maguire will be representing both Mr. Finley and Mr. Grettenberg in the drawing of the contract and Mr. Campbell can well prepare a contract and forward it to Mr. Maguire—so I think your whole consummation of any purchases you wish to make will be worked through Mr. Maguire.

I have only one additional item that I did not discuss with you that Steve and I have been talking about and that is the desirability of three large warehouses to be placed where you have the three universal buildings. We would suggest warehouses almost exactly identical with the big warehouse being constructed by Mr. Grettenberg while you were here. The only change we would suggest would be a door on each end that will let implements go into it.

I have been bold enough to suggest to Mr. Grettenberg that he obtain prices on all of the structural material F.O.B. the port, thinking that the steel companies have spent millions of dollars arriving at the best and cheapest construction—and that three or four models should be available

2. President of Cambell Dryer Company, in Des Moines.

for the people in your steel industry to study as typical of the most modern American design of warehousing facilities.

I visited with Mr. Grettenberg this morning and he is perfectly willing to send a mechanic to arrive in the Soviet Union at the same time the machinery arrives and to help with the assembling and demonstrating of each and every one of the pieces of machinery being sent over—and of holding a school for the mechanics in each area. Mr. Grettenberg, himself, is willing to come over, and spend a limited time—two or three weeks—helping organize the supply of extra parts and constructing knowledge on the best methods of handling repairs, etc.

And now we come to the only remaining part of putting on what I believe is a magnificent demonstation—and that part is my part—the seed business.

As I explained in my letter to you when you arrived, Garst & Thomas are not in a position to accept an order for seed at this time. Our crop is not advanced far enough—either our crop of hybrid seed corn or our crop of Garst & Thomas hybrid grain sorghums.

I do know this—that you will plant 20 million hectares of corn next spring, half for grain and half for silage. That means you will plant 10 million hectares for silage—and I know that the bulk of the Garst & Thomas corn for the bulk of the planting in Russia will be better for silage than for grain. I will tell you this without any hesitancy, however—I think there is as great or greater profit in the planting of hybrid corn for silage as there is for grain. The proportion of grain in the silage is higher, the stiffness of stalk makes the ensiling of hybrid corn much faster and easier, and the fact that the ears have a tendency to ripen with hybrid corn faster than the leaves lets the resultant product make better ensilage.

I am hopeful and expectant that Garst & Thomas will be able to accept an order for as much as three thousand tons of corn directly after the harvesting season has been finished. I am also hopeful and expectant that Garst & Thomas can accept an order for up to one thousand tons of Garst & Thomas hybrid grain sorghums although it may not be that much that we can supply.

As I have explained repeatedly before, I have always felt that when I sold an American farmer any hybrid seed corn, it was my duty as well as my pleasure to show him how to secure maximum results in yield from that bushel of corn with minimum effort—and I have always felt that I was entitled to show him how to get 100 percent use out of the final crop in the feeding of it to cattle, chickens, and pigs. I have taken this attitude for twenty-five years with American farmers.

I feel no different about Garst & Thomas's foreign customers. I think when we sell corn to the Soviet Union, to Rumania, to Bulgaria—to any customer anywhere—that we are entitled and should be helpful to see that that customer raises the biggest crop and feeds it with the greatest efficiency.

Most hybrid producers in the United States have not taken this broad attitude of helpfulness. Perhaps it is because most producers of hybrid seed do not, themselves, farm. They don't know the problems of farming. They don't know about feeding—they are just unable to be as helpful as we are.

I have had a great feeling in the last several years of having contributed at least somewhat to a lessening of tensions between our two countries and I think increased trade will continue to lessen tensions. It is the only hope of the world, as I see it—not a continuation of the cold war, but a gradual warming of the relations between our two countries and a doing away with of tensions.

The Hungarian uprising was a very bad blow to better relations between our countries. As you are aware, the recent publicity about [Imre] Nagy[3] has had a distressingly bad effect on the general American opinion toward the Soviet Union.

I am anxious to accept Mr. Matskevich's invitation to come and give a series of talks on how to best raise corn and how to most efficiently feed corn in the Soviet Union. I will not be able to come before January of 1959, but I expect to come in January, February, or early March, and I will do so unless the relationships between our two countries worsen rather than improve. But I will not be able to come and will not come if international relations become more tense. I will come with the greatest kind of pleasure if they become less tense. . . .

Sincerely yours,
Roswell Garst

July 24, 1958

From Vladimir Matskevich
[*Unofficial translation*]

I was very glad to hear from Mr. Tereshtenko that the honorary degree of Doctor of Laws had been granted to you by Grinnell College in Iowa. Please, accept my most hearty congratulations on this occasion.

3. After a secret trial Nagy was executed on June 16, 1958.

In my opinion this act of one of the oldest and well known colleges in the USA is the recognition—which you have undoubtedly deserved—of your outstanding accomplishments in the field of effective methods in agriculture in the USA, particularly in corn production.

Your greater interest in recent years towards the international relations in the field of agriculture, which are very important for the establishment of better relations among all the peoples of the world, even more emphasizes your accomplishments.

I am sure that the high degree in science granted to you will further inspire you in your activities and I would like to avail myself of this opportunity to wish you further success in your work.

Respectfully
Vladimir Matskevich

August 19, 1958

To Nikita Khrushchev,

. . .At the invitation of Mr. Matskevich, I am planning on visiting the Soviet Union at my earliest convenience, which will be some time after January 1, to talk with your agricultural leaders about the great opportunities that lie ahead. When I do come, I would like ever so much to visit with you and at some length.

I never let anyone tell me that you Russians want war. And I assure you that we Americans do not want war either. If another world war occurs, civilization as we all know it, disappears, so it is idle to talk about war, wasteful to prepare for war, and equally silly to be preoccupied with "peace" if peace means only "no war."

Yet the world is full of war talk, full of arms, and is apparently thinking of little else than how to survive a war, really knowing that it is impossible.

Would it not be a good idea to forget the whole business and get on with something that would be enjoyable?

If we whip the problems of production of useful things on a worldwide basis, what a happy world it would be.

I do not like the word *coexistence*. To me it means two different camps just agreeing that they won't fight. I don't think that is good enough. I am unwilling mentally to settle for *coexistence*.

I believe a real cooperation is so much better that the world ought not to be willing to settle for less.

The Soviet Union, industrial Western Europe, and North America—not only the United States, but Canada included—this whole great area has already developed productive ability to give the whole area an increasingly high standard of living. This whole area could cooperate in furnishing capital and knowhow to the rest of the world to rather rapidly raise the standard of living of the whole world.

I have lived through the period when the ability to produce things has changed the United States from an economy of scarcity to an economy where we do not know what to do with all of the things we can produce.

I have seen your country and also Rumania, and in agriculture, at least, I am sure that you are only a few years behind us. I do not question at all that you are well on the way to abundance in other things.

It took us a long time and a lot of hard work to master the problems of production. Your people, with your splendid educational institutions and with your peoples' willingness to learn and to work, will, in the rather foreseeable future, obtain and enjoy a much higher standard of living.

The most enjoyable activity for an American farmer these days is to show some other farmer in some other part of the world how to produce more food. I am sure from my association with the farmers of the Soviet Union that they are the same as American farmers in this respect. Nor do I think the Russians and Americans are unique in this attitude—person-to-person all people like to be helpful.

I believe that the best way to join together in a cooperative effort to raise the standard of living for all of the peoples of the world is through the United Nations and with a program big enough to inspire the people of the whole world. The Middle East and Asia would be a good place to start—but that would only be a start.

I wish to make one suggestion to you. That suggestion is that you never mention war or threats of war in another speech. Recriminations of any kind are a waste of time.

We all tend to judge ourselves by our ideals and the other fellow by his actions. We all act so far short of our ideals that there is no end to the possibility of charging that the other man is a hypocrite.

Peculiarly enough, the charges and the countercharges between the two governments seem never to be true of the individual citizens of the two countries. Your country's citizens have always treated me with great cordiality in spite of the fact that the two governments charge and countercharge each other with war threats and all sorts of other unfair and unwise and unjust tactics.

I hope the Soviet Union and the United States and Western Europe can set a goal worthwhile for the world as a whole. Let's set out to conquer want and misery on a worldwide basis and put the problems of production of things behind us.

The last ten years, the people on this globe of ours just must have been spending something like $100 billion a year preparing for a war that nobody wants, nobody expects and that no one could survive. On a global basis, that situation is completely insane.

It seems to me that you have the greatest opportunity in world history—that you personally have this opportunity. It will take great courage on your part to be accused of instigating war without you yourself retaliating with the same kind of talk. But if you can accept charges that you believe to be unfounded without retaliating verbally, a great step forward will have been taken.

Likewise, President Eisenhower must have great courage if he is able to stand up against the charges that the government of the United States is provoking war—and not retaliate in kind with charges that it is someone else who is creating warlike situations.

If the *governments* of our two countries could quit the charges and countercharges—the bitternesses—and imitate the citizens of the two countries, this world would be a good deal better place to live right now. If the governments of the two countries could cooperate as well as the citizens of the two countries, very materially reduce the cost of armament, and join together in a cooperative effort to bring a higher living standard to all of the people in the world, what wonderful progress we could make.

These are things I hope to discuss with you personally. These are things I hope I can discuss with President Eisenhower personally.

I have never had any connection with the American government other than that of a private citizen of the United States. I have no political ambitions to ever become connected with the American government other than as a private citizen. I am just a farmer.

The opportunities facing the world are the greatest in history. Productive capacitites in all lines have improved so greatly in the last twenty years that the whole world can now have in the very foreseeable future, not only the thing I wrote about, "food unlimited and good food," but decent transportation, decent medication, decent education, a much, much higher standard of living than was ever possible before.

I think your country and my country should not live a life of coexis-

tence. I think they should live a life of cooperation, not only for the benefit of both countries but for the benefit of the peoples of the world.
I refuse to settle mentally for less.

Very sincerely yours,
Roswell Garst

February 8, 1959

To Mr. N.S. Khrushchev
The Kremlin
Moscow, USSR

Dear Mr. Khrushchev,
Mr. Mikoyan has probably been able to tell you about the fine visit I had with him in the Soviet Embassy at Washington, D.C., while he was in the United States.

As I told Mr. Mikoyan, I am extremely disappointed that a really good demonstration program such as I suggested to your several delegates cannot now be put into practice in the Soviet Union in the year 1959. Let me tell you the situation as I see it.

In June, Mr. Emelianov, Mr. Gitalov, Mr. Shuydko, and Marina arrived in the United States. I took them immediately to my farm. I let Gitalov and Shuydko plant some grain sorghums with a four-row planter. I let them cultivate crops with four-row cultivators and six-row cultivators. I let them chop hay. I let them feed cattle. I let them make green mass from lucerne; I showed them our feed-mixing plant, our drying plant, our dump wagons; let them actually work out on the farm; and did the best possible job of showing them that although one man could not farm one hundred hectares, six men could farm eight hundred hectares with ease.

I showed them the most excellent demonstrations of fertilized corn as against unfertilized corn. I showed them demonstrations of corn which had had proper insecticides and herbicides used as opposed to corn which had not had insecticides and herbicides used. I showed them our methods of farming and let them partake in our methods of farming.

Later during the summer my farm was visited by Mr. Chekmenov, the deputy minister of agriculture to Mr. Matskevich, and by Mr. [Mark] Spivak, the president of agriculture of the Ukraine, and Mr. Posnitny of the collectivist farm down near Odessa, and by numerous other members of your agricultural delegations that were visiting the United States. My son

Stephen and my son David, and my nephew, Mr. John Chrystal, left no stone unturned to show all of the people you sent over most completely how we do farm.

I was asked to make recommendations as to how I thought it would be best for your country to get a look at American methods.

With the assistance of my two sons and with the assistance of my nephew, John Chrystal, and with the assistance of Mr. Grettenberg of the Grettenberg Implement Company here in Coon Rapids, we spent a good many hours putting together exactly what we would use on units of eight hundred hectares of corn, and we recommended to your delegations the establishment of six units within the Soviet Union of eight hundred hectares each for the demonstration of the fact that six men could cultivate 800 hectares of corn.

We recommended feed grinding and grain drying establishments with what we call a "universal building" such as I have in our farm at Coon Rapids on three of these units, and we recommended the purchase of portable grinders for at least three other units. We recommended the purchase of the best type of American fertilizers, of insecticides, of herbicides; and of course, we recommended the use of Pioneer hybrid seed corn. I recommended the purchase of at least several large warehouses in connection with these demonstration units. I think the total recommendations that I made to the several persons must have approached a figure of something like three million dollars. . . .

I was, in fact, ashamed to recommend such a small amount for a country which has such a very large opportunity to increase its agricultural productiveness and, at the same time, materially reduce the labor requirements of agriculture.

As I told Mr. Mikoyan when he was here, I should have recommended, for instance, in addition to what I did recommend, machinery for planting, spraying, and digging of potatoes as is done by the most advanced American methods. When I was in the Soviet Union I saw as many as three or four hundred people in one potato field digging potatoes by fork. No American potato grower could possibly afford hand labor for the digging of potatoes, and it is my considered judgment that you cannot afford such a wasteful method of raising potatoes. As I told Mr. Mikoyan when he was here, equally suitable machinery is available for the planting, cultivating, and harvesting of sugar beets. I did not make nearly as many widespread recommendations as I feel I should have made.

Furthermore, I stressed to all of the people who visited here that I wanted my recommendations accepted in toto or not accepted at all because I did not want to be connected with the failure that is sure to come from the partial acceptance of broad recommendations. I was told that 85 percent of the corn that is grown for the harvest of grain in the Soviet Union is now planted with hybrid seed but that you had not, as yet, used hybrid corn for fodder. In the United States, the use of hybrid corn was, in the early stages, much more popular for fodder than it was for grain because in the case of hybrid corns, the stalk and the leaves stayed green after the grain had started to dent and had begun to ripen and the fodder produced from the use of hybrid seed corn is materially better than the fodder produced from varieties of grain. I, therefore, recommended the widespread sampling of hybrid corn for fodder purposes throughout your whole area. I was informed that out of some 20 million hectares of corn that will be grown in the Soviet Union this next year, approximately half is used for grain and approximately half is used for silage. Because of my positive knowledge of the virtues and superiority of hybrid corn for fodder purposes, I recommended the purchase of three thousand tons of Pioneer hybrid seed corn for demonsration purposes.

And I urged upon every one of the men involved the necessity of an early decision in these matters so that if they decided to go ahead with their recommendations, the things could be shipped in the fall of the year and arrive in time. When I heard nothing from them in October, I started wiring Mr. Emelianov, Mr. Matskevich, and others, urging the importance of an early decision. In mid-December, I even wired you and received an acknowledgment of my wire to you, but then I did not hear anything from them until I got a letter today from Exportkhleb ordering very small and experimental quantities of Pioneer hybrid seed corn and Garst & Thomas hybrid grain sorghums—ordering them so late that there is no chance that we could get the export license and get them shipped to the coast and get them over there in time for proper planting.

And Mr. Grettenberg got an inquiry about some machinery here in February when it is impossible to get any machinery for export purposes because the International Harvester Company and the other suppliers of machinery are so busy taking care of the large domestic demand that they will not interfere with their scheduled shipments to American customers.

As I told Mr. Mikoyan, I think that bureaucracy and the lack of arriving at decisions has made it impossible to have any good demonstrations within the Soviet Union of the American methods which I recommended during the year 1959. It makes me heartsick because I thought it could be so helpful to your country.

I told Mr. Mikoyan that never again was I going to graciously entertain visitors who came to look and to get advice but not to act. I am perfectly willing to contribute all of the accumulated knowledge I have. I am perfectly willing to contribute my time and my energy and my thoughts to the helping of you in your promotion of greater agricultural productivity in the Soviet Union if you are going to accept that advice and act upon it. But I am impatient with the fact that your government, after studying my recommendations, divided my recommendations by ten, whereas I thought they should have been multiplied by ten.

My father used to use an expression in describing people that I have always liked. If he wanted to describe a man who had small ideas, he used to say of him "He would take two bites at a cherry." He taught me to look down upon people who would take two bites at a cherry and I always have done so.

I have never thought of the Soviet Union as being a nation that does little things—I think of you as a great nation that thinks in a big way. And yet it seems to me that so far as agricultural programs, your nation is "taking two bites at a cherry". . . .

> Very sincerely yours,
> Roswell Garst

May 1, 1959

To Ambassador Mikhail A. Menshikov
Washington, D.C.

Dear Ambassador Menshikov,

As you know, Mrs. Garst and I have just returned from a Mediterranean vacation and a visit to the Soviet Union, Hungary, Rumania, Bulgaria, and Yugoslavia. . . .

I want this letter to remind Mr. Khrushchev of what I told him when I visited with him at Sochi. That is, I hope he never again speaks of the power of the Soviet Army or the power of the intercontinental missiles, nor the power of any warlike activity. I think when he talks about these things it leads our own military men to talk about the same thing.

I suggest to Mr. Khrushchev again via this letter that the thing he and the Soviet Union should be talking about is the insanity of this world spending $100 billion a year preparing for a war that nobody wants, nobody expects, and nobody could survive.

If Mr. Khrushchev talks about the insanity of the whole world spending this terrific amount of money wastefully—and talks about nothing else—the United States will be forced to talk about the same thing. If Mr. Khrushchev talks about the insanity of the armaments race, he will of course have to think about the insanity of the armaments race. And if he talks about the insanity of the armaments race and we are forced to talk about the insanity of the armaments race, we will be forced equally to think about it and stress it.

What a wonderful world we could all have, what progress the whole world could make, what a wonderful lift of the living standards of all of the people of the world could come if we could cut the armaments burden in two—and find out that worked so well that we cut it in two again.

I could, of course, have forwarded this letter directly to Mr. Khrushchev for interpretation in Moscow. I thought however, that you would enjoy knowing my thoughts—and I thought it might be valuable for you to know what I did suggest to Mr. Khrushchev during my visit with him.

Very sincerely yours,

Roswell Garst

August 8, 1959

To Vice President Nixon
Senate Office Building
Washington, D.C.

Dear Mr. Vice President,

First I want to congratulate you on having done an extraordinarily good job in the Soviet Union and in Poland with your visit. And I want particularly to congratulate you on the talk you gave at the airport upon your return—the talk in which you urged Americans to be courteous to Mr. Khrushchev when he arrives in the United States.

You undoubtedly know that he announced at his press conference in Moscow that after he had visited with President Eisenhower and visited New York and Washington he wanted to visit the Garst Farm at Coon Rapids, Iowa.

I believe there is a logical conclusion that may be drawn from the fact that the only thing he mentioned other than seeing the president and Washington and New York was the Garst Farm. I believe that it indicates quite clearly a very high interest in agriculture, and I believe that this very high interest in agriculture is logical and sound and of almost primary importance to them.

They are getting on top of their housing situation. They don't do it as well as we, of course—but they are getting their people housed. I suspect that they come closer to the United States in mining and industrial production than they do in agriculture. I do know that when I went there in 1955 agriculture had been neglected ever since the war—probably wisely neglected—in favor of recovery from the war damage, housing, and some expansion of utterly necessary industries. They just had other things that were of much more importance than agriculture. They could eat a good deal of bread, a good deal of borscht and get by on a poor diet but still survive.

And a rather poor diet they certainly did have at that time. It is interesting to note that the first delegation they sent to the United States in 1955 was an agricultural delegation—further proof of their decision to emphasize their need of an expanded agriculture.

When I was there in the fall of 1955, I explained to everyone that I realized that they had been neglecting agricultural production in favor of even more necessary improvements, that I knew they had to rebuild Stalingrad and the other cities, that I knew they had to emphasize housing, that I knew they had to rebuild their factories, that I did recognize that they had neglected agriculture. . . . .

Now comes Mr. Khrushchev to visit in Iowa. What a sight he will see. Iowa, according to the government predictions, will raise this year more than 720 million bushels of corn on 56,000 square miles—an average of more than 12,000 bushels per square mile, a full 10 percent of the expected global corn production—and Mr. Khrushchev is interested in corn and the products of corn. And Iowa is going to raise this crop with relatively less labor than has ever before been spent upon such a vast amount of food. Iowa is going to transfer this terrific corn crop largely into pork and beef and milk and eggs—the very products that are in short supply in the Soviet Union.

It seems to me that Mr. Khrushchev's visit to Coon Rapids sets the stage for a speech by either President Eisenhower or yourself and no one of lesser stature in our political scheme. Khrushchev's visit to the United States will be fully covered by every news facility—newspapers, television, radio, and magazines. Even behind the iron curtain, where we are most anxious to have people read and know the true situation of this world, the Khrushchev journey will be highly publicized and well covered.

I think after we have shown Mr. Khrushchev some hundred-bushel corn fields—one hundred bushels per acre or more—after we have shown him high mechanization; after we have shown him how to scientifically transfer corn into pork, eggs, beef; after we have given him a good display of American efficiency in agriculture; and after we have given him time to himself, he will express the thought that this was what he wanted to see, the thought that this was what he wants to see come about within the Soviet Union, which things I am sure he will say. Then I think we have a perfect setting for a talk by the president or by yourself.

Either of you could review the contribution that the American farmers made to the actual winning of World War II. You could review the vast amounts of food we sent just at the end of the war—including a very large amount of food to the Soviet Union itself. You could review the amount of food that has been exported under either outright grant or under Public Law 480 since the war. You could point out that we have been extremely helpful to the hungry nations of the world—and that we expect to continue to be helpful to the hungry people of the world.

There has been a good deal of question[ing], according to press reports, whether we ever got credit for our shipments of food to the Soviet Union and the other countries at the end of the war, whether we ever got credit for the food shipments to India and to other countries that we sent a lot of food to under the Marshall Plan. The total must now be rather staggering. And the total on hand is rather staggering—that we can and will ship.

It certainly would be a fine place to congratulate American farmers on the terrific job they have done. We have 28 million more people now than we had in 1950—we will have an increase in population of 20 percent between 1950 and 1960—and yet American farmers have not only improved the diet during that ten year period; they have been able to produce enough to not only feed the extra people and feed them better; they have produced enough to continue to help the hungry nations. They should be congratulated—and strongly congratulated.

And certainly the American consumer should be told that the best diet in the world costs him less than 25 percent of his total wages—a phenomenally low percentage of his total wages. This point should be made very strongly because in many parts of the world—and specifically behind the iron curtain—it takes well over half of the total earnings to buy the food alone. And in some countries, it takes 75 percent of the total earnings to buy the food. I think a paragraph should be worked in pointing out to the American consumer that the surpluses we have produced have reduced his food bill to a very much greater extent than his share of the taxes necessary to support the farm program, which is *literally true*.

The Census Bureau predicts a gain of 40 million people during the 1960s. We are fortunate to have agricultural production up to its present level—and our present level of production must be maintained—or food costs will go up very distinctly in the next decade. We probably will not be able to contribute ten years from now as generously as we have been contributing in the past ten years—but we can do so at the start.

I do not believe that the secretary of agriculture of the United States, Mr. [Ezra Taft] Benson, ought to be considered for this talk, although that might appear logical. I think this talk definitely ought to be made and I think it ought to be made either by the president or by you. I think it's important to have it widely quoted all over the world—and I think it ought to be given by the president, who is in charge of our foreign policy, or by the vice president, who probably will in future be in charge of our foreign policy, and by no one of less stature.

I expect to invite the editors of all the Midwest farm papers to Coon Rapids while Mr. Khrushchev is here so that such a talk would have the finest kind of coverage in the whole Corn Belt. . . .

Very sincerely yours,
Roswell Garst

September 9, 1959

To Mr. N.S. Khrushchev
Soviet Embassy
Washington, D.C.

Dear Mr. Khrushchev,

First I want you to know how highly pleased I am that you are coming out to see our farming operation and also how highly pleased I am that Mrs. Khrushchev and your family have accompanied you and will accompany you.

Arrangements will be pretty well completed by Mr. Menshikov and our State Department before your arrival with one exception. Because I think you enjoy a good laugh, and because I enjoy a good laugh, and because I think there is an opportunity for a good laugh, I have a suggestion to make.

According to the plan, the festivities of Des Moines the afternoon and evening, Tuesday, September 22nd, terminate at 10 P.M..

And according to the plans, you must leave my farm by 2:50 P.M. or 3:00 P.M. at the very latest in order to get to Iowa State College for a too-brief appearance in order to get to the airport at Des Moines in time to take off for Pittsburgh.

The only elastic thing at all is when we start in the morning from Des Moines and that is entirely up to you—entirely and completely. I have many things I am extremely anxious to show you. The State Department guessed that you would want to start about 9:00 o'clock from Des Moines. I think that is about the earliest anybody connected with the State Department ever got up in his life. Furthermore, the press don't get very many good stories until after noon so the radio people, television people, reporters of every kind habitually sleep late—and work late. Most people in cities, most diplomats, and certainly most news people think that 9:00 o'clock is just the crack of dawn.

I suggested to them that I thought you were anxious to see a great many things and that you might possibly be able to be ready to leave the hotel at 7:30 in the morning. They were utterly horrified—pointed out that this hour would horrify the press. So on the way home, I thought of a joke that I think you would enjoy as much as I would enjoy.

It gets light here even on September 23 at about 6:00 A.M.—it will be perfectly daylight by 6:30 A.M.. If you say that you want to see a good many things in Coon Rapids—that you must leave by 3:00 P.M. in order to get to Ames, but you see no reason for not getting up in the morning as soon as it is light enough to see—everyone will be, of course, required to do the same thing.

Mr. Lodge will have too get up early, I suspect for the first time in many years. So would every reporter in the United States. You would only be getting up at the same time that every farmer in the United States is getting up—you would only be getting up at the same time all your farm delegates like Gitalov and Shuydko did get up all the time they were here—at dawn.

Now, if you are tired, I still think it would be just as well to get everybody out of bed early once—even though you take a rest of a half an hour or an hour before lunch—in fact, I think you might well want to do that.

I feel certain that Mrs. Khrushchev will not want to see the farming operation and I thank it's a little unfair to expect Mrs. Lodge and Mrs. Khrushchev and the other ladies to get up early. It takes an hour and a half to get from Des Moines to Coon Rapids. If the ladies leave Des Moines by 7:45 or 8:00, that will be ample time because they will come to Coon Rapids directly and look over our little town while the rest of us are looking over the farming operation.

I am not making this suggestion to a single soul other than to you. I am sending it, of course, to Mr. Ambassador Menshikov for translation. I have told my attorney, Mr. Maguire, privately, so he could discuss the thing with

Ambassador Menshikov. I think it ought to be kept a deep secret and not announced by you until the dinner party at the hotel in Des Moines the evening of September 22—the day before you are to come out here.

It will be a fair joke if you even are willing to leave there at 7:30 in the morning, a decidedly good joke if you are willing to leave at 7:00 A.M.; it will be the best in history if you say that you want to leave at daybreak—6:15 or 6:30 A.M.. I'll positively agree that for every half hour you leave Des Moines before 7:30, you will get a half an hour's rest if you want one, before lunch. I can show you a very great many things that will be of very high interest.

Incidentally, I think you will be interested in knowing that the press section of the State Department and the protocol section of the State Department came out the other day for a look at what I was going to show you and they agree with what the radio, television, and newsmen have been telling them—that this will be the highlight of your trip; that it will be the heaviest covered by press, radio, and television of any part of the trip; and that they think it will be your happiest experience on the trip. Mrs. Garst will take very excellent care of Mrs. Khrushchev.

I presume your son and your son-in-law will both want to accompany you looking at the farms on the day you spend here. I presume that Mrs. Khrushchev and Mrs. Lodge and the other important persons will not want to see the farms.

I have told no one except Mr. Maguire that I think this is about the only chance that your two daughters might have to escape and really see the United States as we live. I have two daughters—very delightful daughters—who know every one of the delegates that have ever come from the Soviet Union. They are Antonia, age twenty-six, and Mary, age twenty-three. Both are widely traveled across Europe other than the Soviet Union. Both have been in Rumania for rather decent periods of time. Both of them are thoroughly delightful—and thoroughly experienced.

I think it is possible that if they stayed at the hotel until we had left and the ladies of the party had left that your two daughters and my two daughters along with a driver and security man and an interpreter could follow your footsteps until noon—or Mrs. Khrushchev's footsteps until noon—and do it in almost utter privacy. The big press coverage will be on you and the second largest on Mrs. Khrushchev. Practically the whole press will be following one or the other.

If this plan is in no way ever mentioned to the press—and if only a minimum number of people know it in your entourage or in our American entourage—they could really get a good look at how Americans live and farm and operate. Give this a little thought. I think it would be a wonderful opportunity for your two daughters—and also, of course, for my own two daughters.

I look forward with high anticipation to Wednesday, the twenty-third day of September 1959. I am sure it will be difficult because of the extremely heavy news coverage—but I think it will be thoroughly enjoyable for you and for Mrs. Khrushchev—and for your daughters. We will leave no stone unturned to have it a pleasant day in your memories.[1]

Very sincerely yours,
Roswell Garst

COON RAPIDS, IOWA
SEPTEMBER 24, 1959

TELEGRAM TO VALERY TERESHTENKO
METROPOL HOTEL,
MOSCOW, USSR

YESTERDAY MR. AND MRS. KHRUSHCHEV AND THEIR FAMILY AND THE WHOLE OFFICIAL PARTY SPENT THE DAY WITH US HERE AT COON RAPIDS STOP WE HAD A TENT IN THE YARD JUST AS WE HAD FOR THE WEDDING LAST YEAR ONLY LARGER—IT WAS A DELIGHTFUL LUNCH FOR THE WHOLE OFFICIAL PARTY WITH A GREAT MANY GUESTS FROM THE LOCAL COMMUNITY INCLUDING ALL OF THE PEOPLE YOU KNOW IN COON RAPIDS STOP OUR DEEPEST REGRET WAS THAT YOU AND MR. MATSKE-VICH AND MR. EMELIANOV AND MR. GITALOV AND MARINA AND THE OTHER DELEGATES THAT HAVE VISITED THE UNITED STATES WERE NOT ABLE TO BE WITH US STOP
WILL YOU KINDLY EXTEND OUR GREETINGS TO THOSE PEOPLE IN THE SOVIET UNION THAT WERE NOT ABLE TO BE WITH US YESTERDAY STOP ALSO TELL MR. MATSKEVICH THAT I WILL WRITE HIM THIS WEEKEND FULLY ABOUT THE CATTLE STOP TELL MR. MATSKEVICH ALSO THAT MACHINERY ORDER HAS

1. Khrushchev was amused by Garst's "absolutely unrealistic" proposal and commented on it in his memoirs: "If I tried to run off secretly with Garst when he came to fetch me," Khrushchev wrote, "it might appear that I'd been kidnapped, like a bride in the Caucasus or in Central Asia." *Khrushchev Remembers: The Last Testament,* translated and edited by Strobe Talbott (Boston: Little, Brown and Co., 1974), p. 397.

BEEN PLACED BY AMTORG AND THAT I WILL WRITE HIM ABOUT
FERTILIZER, HERBICIDES, INSECTICIDES, ETC., WHICH SHOULD
ACCOMPANY THE MACHINERY STOP. . . .

                                        ROSWELL GARST

                                        September 28, 1959
To Ambassador Menshikov
Embassy of the USSR
16th Street
Washington, D.C.

Dear Mr. Ambassador:

During the automobile ride from Des Moines to Coon Rapids, I had
quite a long visit with Chairman Khrushchev about what I broadly called
"Cases of Compassion." I told him I thought that it was of the utmost
importance to get these matters taken care of at the earliest possible
moment.

They involve American citizens with relatives in the Soviet Union,
and quite a few of them are most distressing—so distressing in fact that
Mrs. Garst and I have been broken up and saddened by the situations. We
received a surprisingly few of these letters when you consider the great
amount of publicity connected with his trip to Coon Rapids and to our farm.

I pointed out to Mr. Khrushchev that the number of cases of this kind
in the whole United States was not large—and I urged upon him the
advisability of getting them cleared at the earliest possible moment, and he
agreed that they should be taken care of at the earliest possible moment.
He told me to take the matter up with you and through you with Mr.
[Andrei] Gromyko.

I am absolutely certain that all of these compassion cases should be
cleared up just as soon as possible from the standpoint of better relations
between your government and ours. And I will personally appreciate a
letter from you saying that you are giving this matter very prompt attention.

I have made photostatic copies of the letters for my own files and I am
forwarding you the originals. I can only urge promptness.

                                        Very sincerely yours,
                                        Roswell Garst

December 16, 1959

To Chairman Nikita S. Khrushchev
and Family
Moscow, USSR

Dear Mr. and Mrs. Khrushchev and Family,

I am forwarding to you through the Embassy of the USSR in Washington, a couple of presents from the whole Roswell Garst family. One is a bound volume of still pictures covering your trip here to Coon Rapids. And enclosed with it is a sixteen millimeter movie of your trip here to Coon Rapids.

In each case, some of the pictures were taken before your arrival at Coon Rapids when better and more unobstructed views of what you were going to see were available. It may interest you all to know that Mrs. Garst and I have duplicate copies in our own home of both the book we are sending you and the film we are sending and we shall cherish them always.

There can be no doubt at all that your trip to the United States did probably more than anything else could have done to lower the suspicions and the misconceptions that did exist between our two countries. And I am completely certain that having Mrs. Khrushchev and the two Khrushchev daughters and the Khrushchev son and son-in-law in the party made it infinitely more effective than had only you, Chairman Khrushchev, come by yourself.

I have been kept busy as has Mrs. Garst since your visit here, giving talks before a great many groups of rather prominent people. I have pointed out that you had many misconceptions about the United States four years ago and that we had an at least equal number of misconceptions about the Soviet Union four years ago. We have both pointed out that the ever-increasing contacts between the two countries have greatly lowered the misconceptions and the mistrust between the two countries and that the whole situation is much more hopeful than it was four years ago. The greatest kind of progress has been made and is being made.

I am sure that the visit of President Eisenhower and his family to the Soviet Union next year will add to the better feelings that are growing between our two countries.[1]

The cold war period was a bitter period and a handicap to the whole world. We are all most happy to see it disappearing and conditions warming up.

1. The trip never materialized.

I think it must be replaced finally by sincere friendship. I do not expect that to be in the immediate future—but I think it should be the ultimate goal.

I simply have to repeat what I told you, Mr. Khrushchev, when Mrs. Garst and I saw you last spring. It is insanity for the world to spend $100 billion a year preparing for a war that nobody wants, nobody expects and that nobody could survive.

The whole Garst family—in fact, I think, the whole world—look forward hopefully to the time when the burden of armaments can be either wiped out entirely or very, very materially reduced.

It will, of course, call for inspection of disarmament—mutual inspection—so that each country has complete confidence that disarmament is being carried out as agreed. This mutual inspection is as important to the Soviet Union as it is to the United States.

We are all farmers—all highly interested in agriculture—and not capable of judging in detail what ought to be done diplomatically. But we are full of hope that the diplomats of our country and your country will be able to arrive at a mutually satisfactory basis for disarmament, and we are bold enough to suggest that the United Nations itself probably should be the principal organization involved.

With something approaching a billion hungry people in the world now and with something like 50 million new and additional people in the world each year, one of the world's major problems is going to be the increased production of food on a really worldwide basis. And that's what the whole Garst family is trained to do and love to do—to help people learn to raise more food with less labor.

We hope that the agriculture of the Soviet Union exceeds the goals of the Seven Year Plan in agricultural production. We think your agriculture has terrific opportunities for increased production with less and less labor.

We were all proud to have you visit us. With the season's greetings from all of us to all of you, we are

Sincerely yours,
Roswell Garst

December 21, 1959

To Chairman Nikita S. Khrushchev,
Minister of Agriculture Matskevich
The Kremlin
Moscow, USSR

Gentlemen:

Today I wish to write you strictly an agricultural letter. I hope it will arrive about the time 1960 arrives so I want to include in it my best wishes for agricultural progress in the Soviet Union in 1960—and in the years ahead.

The farm machinery ordered through Amtorg will be shipped in January. Mr. [Griguri] Volkov[1] says that you are intending to buy some beef cattle, and I understand that you are planning a good big expansion in both chickens for broiler purposes and chickens for egg purposes. All of these matters please me no end—all represent progress; all make me happy.

But perhaps most important, agriculturally, was the recent visit of the chemical delegation who came under the auspices of your minister of chemistry. Fortunately I was able to get my brother, Jonathan Garst, to help them take a careful look at our fertilizer industry and at the production of urea, not only for fertilizer but for the feeding of ruminants.

Protein is of worldwide shortage. We do not even have enough in the United States for optimum rations of our livestock and poultry. The one unlimited source of protein is urea—as unlimited as the air, the natural gas, and the water from which it is made.

By using urea as a protein for your ruminants—your cattle and sheep and goats—you can steal from them the by-product meals such as cottonseed meal, flaxseed meal, sunflower meal, bean meals, and the rest and use those by-product meals for your ruminants—for your nonruminants, it should have been: for your chickens and your pigs—and you can learn to supply all of the protein of the ruminants in the form of urea. . . .

I just wanted to write this letter to both of you because I think it is occasionally easy for people to forget that the chemical industry has made very, very substantial contributions to the increased productiveness of American agriculture.

Similarly, the chemical industry can make very, very substantial contributions to Soviet agriculture. It is now estimated that a full one-third of American agricultural production is attributable to chemistry in the form of fertilizers, herbicides, and insecticides.

1. Agricultural counselor, Soviet Embassy, Washington, D.C.

In the case of the Garst family, we are inclined to think that as much as one-half of our total production is due to chemistry. By using fertilizer and insecticides, we are able to grow corn after corn after corn on that land which gives itself best to corn growing. We are able to have pasture after pasture after pasture with ever-increasing grass crops. We are now pasturing more than twice as many cattle on any area as we did pasture before fertilizing. . . .

I was highly gratified that the summit meeting has been set up for next spring and even more gratified with the suggestion that there be not one summit meeting but continuing summit meetings. That looks sensible to me—and hopeful.

May 1960 bring to you and your country continued progress in the production of more food and fiber with less labor. That is my sincere wish.

Sincerely yours,
Roswell Garst

January 26, 1960
To Chairman Nikita Khrushchev
The Kremlin, Moscow, USSR

Dear Chairman Khrushchev.

First I want to thank you for the very beautiful album and the original letter from you containing the New Year's greetings from you and your family to me and my family. I had already had before New Years, an English translation of the letter, which has given me high satisfaction and great pleasure. Also, I have had copies of both *Pravda* and *Izvestiia* containing the copies of my New Year's greetings to you and yours—and your greetings to me and my family.

Today I have written a rather long letter to Minister of Agriculture Matskevich, a copy of which is attached hereto. I am forwarding it through your embassy in Washington—through Ambassador Menshikov, suggesting to him that perhaps it would be more convenient for you if he furnished you with the original and also his interpretation into your language. . . .

Kindly note that I have suggested to Mr. Matskevich that he should invite my nephew, Mr. John Chrystal, to visit the Soviet Union from late June until late August of 1960 because I think he could be of extreme benefit to your agriculture. He is the brightest farmer I know—he knows as much about agriculture as I; he entertained Gitalov and that delegation for several months and is a grand person. I suggested to Mr. Matskevich that the invitation be issued without publicity to avoid the press and the photographers.

I am completely certain, as I told Mr. Matskevich, that you and he must "cut the Gordian knot" of bureaucracy in order to accomplish the objective of agricultural expansion as outlined in your Seven Year Plan. I am sure that decisions must be arrived at promptly and executed promptly.

I think you and Mr. Matskevich should study the recommendations I have made. I am quite sure you will broadly agree with the recommendations. I should be most pleased to get your reaction as well as that of Mr. Matskevich.

Nothing could give me greater pleasure than to see Soviet agricultural production zoom upward at an even greater rate than the goals outlined in your Seven Year Plan. I believe it is perfectly feasible to exceed the goals.

Remember that chemistry contributes a full third to the productiveness of the United States agriculture and that Mr. [Georgi] Uvarov, the minister of chemistry, has an extremely important effect on agricultural production.

What progress the world has made since I met you in the fall of '55 in better understanding. May that progress be accelerated even more rapidly from now on. That is my sincere hope, and the hope of my whole family, and, I think, the hope of almost all people in the whole world.

Very sincerely yours,
Roswell Garst

January 26, 1960

To Mr. Vladimir Matskevich,
Minister of Agriculture of the USSR
The Kremlin, Moscow, USSR

Dear Mr. Minister,

I had to go to the hospital January 4 for surgery—I had my gallbladder and appendix removed—so I have been out of circulation for about three weeks. I am practically fully recovered and look forward to even better health than I have enjoyed in the past. While I was convalescing, I had a great deal of time for thought and my thoughts concentrated on how we of the Garst family could be helpful to you people in the Soviet Union in increasing agricultural production and decreasing the amount of labor.

I have some very specific suggestions. I will try to keep this letter as brief as possible, as brief as is consistent with a good explanation; that does not mean it will be brief, because it is an extraordinarily large subject with which we are dealing. Because of his high interest in agricultural matters, I am sending a copy of it to Chairman Khrushchev. . . .

The first thing I would do would be to get the urea plant contracted for at the earliest possible moment . . . . You should also immediately contract for facilities for the manufacturing of insecticides like Aldrin and herbicides like Randox. Such facilities will be much less costly than the urea plant—and utterly necessary for modern farming.

You should, in my opinion, place another order for machinery at least as large, if not twice as large, as the order that will be shipped three weeks from now. It was just too small an amount to give you the optimum amount to study, or so I believe.

You should conclude your orders being discussed with Mr. Finley of the Finco Company for full and complete equipment for the production of eggs and broilers for at least a few well organized poultry production units and for at least some of the dairy cattle installations that you are contemplating. I have no doubt that you will proceed with warehousing and drying equipment which Mr. Volkov is presently discussing with my son, Stephen.

And you should arrive at a decision about the beef cattle and place that order immediately. Actually, I think I will write you another letter about the beef cattle separate from this within the next few days.

I am completely certain that the Pioneer Hi-Bred Corn Company now has hybrid egg laying strains of chickens that are phenomenally better than even the original chickens you bought from them—and that you should be buying foundation stock from them for the production of your own egg layers. If you wish me to do so, I can take the matter up with them as to how many they could supply and at what time of year. But I am completely certain that this recommendation should be given your immediate attention.

There are only certain times of the year when I can leave Coon Rapids or when I can pay any great amount of attention to helping you with your problems. I do, as you know, have a very large and prosperous and growing business of my own here in the United States which requires my full concentration and effort most of the year.

But I have a nephew Mr. John Chrystal, whom I want to recommend to both you and Mr. Khrushchev. He is thirty-five years old and unmarried and can therefore, get away from home. He is the man who took Mr.

Gitalov, Mr. Shuydko, Mr. Emelianov, and Marina [Rytova] and educated them so thoroughly in the summer of 1958 when they spent several months in the United States. Gitalov actually lived with Chrystal—planted corn for him, cultivated—helped harvest it. He is, in my opinion, the best informed and deepest thinker available for coming to the Soviet Union and making a good, deep study of your problems and ways and means that we could be helpful with advice. . . .

Very respectfully yours,
Roswell Garst

February 24, 1960

To Mr. Valery Tereshtenko,
219 E. Fifth Street,
New York 3, N.Y.

Dear Valery,

I am happy that you are going to the Soviet Union shortly. I hope before you go you will go up and talk very seriously with the people at Amtorg—and that also when you get to Moscow you will get along these same ideas to Mr. Matskevich.

I think no one in the United States has tried to be more helpful in the last five years than have I. As you know thoroughly, I have not been interested in the commercial aspects of our relationship. I have just been trying to promote better understanding between the two countries; I have just been trying to be helpful to the Soviet Union in getting their agriculture really in gear. Many times during the period, I have about given up trying to be helpful because it has been so slow and so difficult and so frustrating—and because my attempts to be helpful have so frequently caused my friends out here an untold amount of work with no results.

Let's take the Grettenberg machinery as an example. As you know, the four delegates of 1958 came in June and went back in October: Emelianov, Gitalov, Marina, and Shuydko. John Chrystal; Steve, and Dave, and I; and Mr. Grettenberg spent hours and days giving them the most complete kind of recommendations on what they should buy. Mr Grettenberg spent endless hours getting prices on all the equipment—we made recommendations on not only machinery but fertilizer, insecticides, potato equipment, etc., etc., etc.

The order should have been placed within a week or two after they returned to Moscow. It was not placed until late this fall—even after I had visited with both Mr. Khrushchev and Mr. Matskevich last March. There was endless haggling about price—and yet the price submitted by Mr. Grettenberg was utterly fair. Then they sent over shipping instructions which Mr. Grettenberg sent to all of the suppliers—and then about three weeks later, they sent a different set of labels, which created great confusion. Then Mr. Kolibovnikov[1], or however you spell the name, complains because the Melos grinders were driven down to New York—which is the way they should have been delivered.

I wrote all the fertilizer companies and the insecticide and the herbicide companies in the fall of 1958 and got prices—and disturbed them no end—thinking that the order would be placed. I heard nothing from the Russians about the order for a year and a half. Then this spring they want a new set of prices—so I get them—and they are based upon immediate shipment when the fertilizer companies had a low demand because it was winter. I could have had immediate shipment a month ago. Now I doubt if I can get immediate shipment. I may not be able to get shipment at all because urea is extremely scarce—certainly it's questionable whether it could arrive in time for use, anyhow—so I have created a lot more work with no results. It's embarrassing to me—and I have declared a dozen times that it isn't going to work.

So it is with the cattle. I get an inquiry, answer the inquiry, think a delegation is coming, don't know whether they are coming, and if I depended upon their announced intentions, the cattle would die of old age before they decided. I could multiply this a dozen times, as you know.

It shortly will not be worthwhile to me. I'll just have to give it up and tell them they are on their own—that I just have neither the time nor the patience to continue exploration of this item, of that item, or the other item; and ask my friends to look up this detail and that detail and the other detail; and then have nothing happen until we have done it two or three times. I am just too busy at other things to waste my time at it.

I have a very large seed corn business and one that is growing. We have a very large farming operation and one that is growing. We have the bank here that takes quite a bit of attention. The visit of Mr. Khrushchev entailed the answering of hundreds and hundreds of letters—and I am getting fairly well caught up on the whole thing. I will continue but only on one basis—that Kolobovnikov desist from minor complaints, and that Matskevich gets through the bureaucracy so that any visiting delegations about

---

1. Probably Griguri Volkov, agricultural counselor for the Soviet Embassy.

the cattle are authorized to proceed with the purchase, the payment, without a lot of little bits of details that don't amount to a damn interfering.

Actually, every delay they have had has caused additional expense. I told them in 1958 that there would be a strike at International Harvester Company and that the price would go up if they didn't order immediately, which it did. I told them this fall that there would be a steel strike which would make deliveries difficult and delayed—and it has. The difficulty of the whole Grettenberg thing is not my fault—it's the Russians' fault. They just are too suspicious.

If I did business the way they do business, I never would get anything done. I am sick and tired and caught up on the bureaucracy. Matskevich has delightful and competent assistants. They could come over here and order the whole thing—in half the time and with half the energy. They could have bought it all on my first recommendations and saved probably as much as 10 percent of the purchase price. And they are making it most difficult to ever get a bargain in the future.

If they have confidence in me—and I think they should have—they should accept my recommendations promptly and without quarreling about minor details. If they don't have confidence in me, I'd rather not see them again. I just am not a haggler. I never have been in my own business and I will not be for them. If I think a thing needs doing, I go ahead and do it and I don't quarrel around about the small details nor the price. I do, of course, try to keep the price fair—and no supplier has ever charged them more than he should have charged them that I have been connected with. . . .

I think it is lack of decision that is the worst of all. If they want cattle, we will supply the cattle—at an entirely reasonable price, quality considered—but we don't want to take three years of indecision about it. If they want insecticides, we will get them the insecticides—but they would be far better advised to just take Steve's recommendations promptly and without change, which is the way it ought to be.

This is the last letter I am ever going to write you or them about these matters. If it gets any worse than it is, I am going to simply tell them that they have to depend upon somebody else for advice—that we are just too busy to be bothered and in no way interested, and that we are forced to give up our attitude of helpfulness simply because we do not have time enough to advise and advise and advise.

I suggested to you and to them several years ago that if they would just take a million dollars or two million dollars or five million dollars and let me allocate it and spend if for them, it would be far their best method. I am sure that is still true. If they want to send over Matskevich's best deputy with the authority to proceed with my advice, that's fine. But this thing of sending unauthorized delegations to go home and report so that the bureaucracy can argue with them—and then take a year arguing with them—there isn't a darned thing to that. Maybe Mr. Matskevich himself should come over here about the time the president goes to Russia. Maybe he should come as my personal guest—and bring Emelianov along to conclude the decisions he makes.

Sincerely,
Roswell Garst

USSR, 1960–68

The shooting-down of an American U-2 reconnaissance aircraft over Soviet territory in May 1960 dashed the diplomatic expectations created by Khrushchev's visit. A confrontation over Berlin in 1961 produced further tension, and the Cuban missile crisis of 1962 brought the USSR and the United States to the brink of conflict. Throughout these unnerving years, Khrushchev maintained his thrust for an agricultural leap forward, and Garst, although disturbed as much as his countrymen by these events, continued to maintain his relationship with the Soviet leader. In 1960, John Chrystal made an extended visit to the Soviet Union on behalf of Garst and thereby established his own reputation in Soviet agricultural circles. In 1962, Garst had to cancel yet another trip, but in 1963, accompanied by Chrystal, he visited the USSR and had further conversations with Khrushchev. Early in 1964, at the request of the Soviet leader, he prepared an article analyzing Soviet agriculture that was published in *Pravda* and *Izvestiia*[1] and cited in detail by Khrushchev at the February 1964 Plenum as a means of eliciting support for further investment in mechanization, agricultural chemicals, and irrigation. However, the Soviet leader's management of his ambitious agricultural campaigns was increasingly being criticized, and for this and other reasons, he was deposed in October 1964. Garst never saw him again.

The agricultural themes of Garst's correspondence during this period focus on the potential of hybridized sorghum for the dry areas of Soviet farmland; increased investment in chemical fertilizers, pesticides, and herbicides; and further development of the general infrastructure of Soviet

1. See Appendix.

agriculture. He laid particular emphasis on the building of a road system, a project Khrushchev agreed to in principle but believed that Russia could not yet afford. Still pursuing his personal concern over the arms race, Garst talked to Khrushchev in 1963 about the continuing negotiations over a nuclear test ban, which was agreed upon by the USSR and the United States in August of that year.

After Khrushchev's removal in 1964, Garst's correspondence with officials in the Soviet Union diminished greatly. He attempted to resurrect a dialogue when Vladimir Matskevich was reinstalled as minister of agriculture in 1965 by the Brezhnev regime and urged Matskevich to visit the United States again. From 1965 through 1968, Garst devoted himself largely to other areas of his domestic and international interests.

November 29, 1960

To Mr. [Vladimir] Matskevich, Minister of Agriculture.
Mr. [A.S.] Shevchenko, Agricultural Adviser of the Chairman
Ministry of Agriculture, Moscow, USSR

Gentlemen:

Mr. Chrystal and I have visited at very great length about the agriculture of the Soviet Union ever since his return from the USSR, from Rumania, and from Bulgaria and Hungary. Because we have had the advantage of seeing the agriculture of all four countries, we have much in common to visit about.

Since my first visit there in 1955, very very great progress has been made. When we compare my first trip in 1955 with Mr. Chrystal's recent trip in 1960, we are both highly gratified with the progress.

For instance, in 1955 there was no hybrid corn being raised in any of the four countries. Now the use of hybrid seed corn is universal in the Soviet Union and nearly so in Rumania and very well started in Hungary and Bulgaria. It has taken you only five years to shift from varieties of corn to the hybrids of corn. It took the United States fifteen years to make a similar shift.

We both believe that this is as it should be—that by eliminating the mistakes we did make, you can telescope the time requirement and reduce it by about one-third—or to one-third—so that you ought to make as much progress in five years as we made in fifteen. We are both delighted to see you making such rapid progress.

In the interests of keeping this letter short enough, we are going to point out the fields in which we believe you can make most rapid progress.

1. You have a rather vast area in the Soviet Union where rainfall is not always adequate for the raising of corn because the rains are scattered and you have rather prolonged periods of drought during the growing season. We have such areas in the United States in western Nebraska, western Kansas, western Oklahoma, and western Texas. In the United States, we use grain sorghums as a crop and we feel that you should get the widest kind of background on grain sorghums at the earliest possible date.

It is only in the last eight years that a practical method for the hybridization of grain sorghums has been discovered and the hybrid seeds of grain sorghums are now almost universally used in the area described, that is, western Nebraska, western Kansas, western Oklahoma, and western Texas. In those four states in 1960, the average yield of grain sorghums was exactly the same as the average yield of corn, and this was true in spite of the fact that the grain sorghums are planted on the drier western half of the state and the corn in the better-watered eastern half of the state—and also in spite of the fact that the corn is always planted in the best land—and the grain sorghums on the thinner land.

Grain sorghums have an ability to wait for the rain—they hibernate in hot weather and dry weather and just wait and wait—and when the rains come, they proceed with their growth but they have not been seriously damaged in the meantime. Without hybridization, grain sorghums were not high yielding. With hybridization, they are extremely high yielding. We would recommend the purchase of a relatively small amount of hybrid grain sorghum seed immediately for planting in the spring of 1961. We were thinking in terms of fifteen thousand fifty-pound bags, and we would suggest that the planting be scattered very widely across the Ukraine and in Uzbekistan and in parts of southern Asia so that you get the widest possible kind of distribution. Both Mr. Chrystal and I ran across high interest and I believe that you will remember that Dean Lambert and the first American farm delegation advised your agriculturists that grain sorghums were a crop that should be widely grown. It takes no change of machinery—and you only use four or five pounds of seed per acre and Garst and Thomas will be happy to sell you the seed for twelve cents a pound F.O.B. port, thoroughly treated with a fungicide and an insecticide and double-bagged for export with a guaranteed germination of 90 percent. It is the finest seed available in the United States.

2. We are completely certain that your hybrid corns and your small grain, such as wheat, oats, barley, and rye, and the grain sorghums; and your truck crops would all yield phenomenally more if you had adequate supplies of fertilizer. Mr. Chrystal says that he has never seen more beautiful wheat in his life than you have even without fertilizer. I was never there in the summertime. But Mr. Chrystal says that in flying over the cornfields, the nitrogen deficiences were very apparent and very severe and that there can be no doubt that the yields of your corn would be doubled with sufficient nitrogen.

For this reason, we are extremely anxious to see you expand your nitrogen facilities and we think that the United States production of nitrogen has made a phenomenal contribution to the crop yields in the United States. My brother, Jonathan Garst, who spent a good deal of time in Rumania and in the Soviet Union looking over your nitrogen fixation capacity and facilities, is absolutely certain that the American nitrogen fixation facilities are the most modern in the world—very simple, very compact, very efficient—and we all feel that you should proceed with the negotiations started [for] a chemical plant from the Chemical Construction Company as per his recommendations.

3. We are definitely sure that you should proceed immediately with the building of facilities for handling your harvest—facilities for cleaning, drying and warehousing. Mr. Chrystal and my son, Stephen, and Mr. George Finley, did a great deal of work last winter planning facilities of this kind that would be as modern as the best in the United States, and Mr. Chrystal himself observed the wish of your agriculturists to be able to harvest at somewhat higher moisture contents than you have in the past to eliminate field loss.

Actually, Mr. Chrystal and I also think you should buy a full set of prefabricated American-type farm buildings made of steel. Those buildings are available for the housing of cattle, chickens, hogs, and machinery; for the warehousing of grain; for practically all purposes. They can be insulated; they are [of] simple, easy construction; and because the USSR now has a very large total steel-producing capacity, we think you ought to have the most modern possible design that has been developed and that is avilable here in the United States.

4. Without any doubt at all in our opinion, you should buy from the Pioneer Hy-Line Chicken Company, Des Moines, Iowa, suffi-

cient foundation stock to greatly expand the number of heavy-producing egg-laying hens. The breeding of hybrid chickens is more advanced in the United States than anywhere else—and we are completely certain that you, from your own experience, know that the Hy-Line will produce far more eggs than ordinary breeds. We think this should be one of the things you do in a rather large way and almost immediately.

5. Both Mr. Chrystal and I know that we do one thing in the United States faster, cheaper, better—and by a very wide margin—than is being done in the Soviet Union, and that is the building of roads. I am not talking about superhighways—I am talking about transportation from the state-owned farms to the major highways and to the towns. We have in the United States the most marvelous road building machinery extant. It can be run by semiskilled mechanics. It consists of very large crawler-type tractors, bulldozers, scoops that are self-loading and very large, and grading equipment that is hauled behind the crawler-type tractors, and dirt elevation equipment. A relatively few men with such machinery can grade several miles a day—it is one of the phenomenal developments of my lifetime. We would recommend the purchase of substantial quantities of such road building equipment of the very latest design.

We are completely certain that the above five recommendations are all of extreme importance and that you should proceed with these purchases at the earliest possible moment—and in rather substantial quantity—so that they could be used in 1961. Certainly the suggested purchase of hybrid grain sorghum seed ought to be almost consummated by cable and immediately; the grain handling equipment ought to be ordered immediately so it could be used in the fall of 1961; in fact, only the nitrogen plant, of the four suggested, could not be used in 1961 if orders were placed immediately. . . .

Amtorg have already shown an interest in the purchase of road building equipment and there has been some resistance in the United States Department of Commerce about issuing export licenses. Apparently the U.S. government has previously classed them as strategic machines that could be used for your military forces. I am pointing out to the Department of Commerce that they are, in reality, utterly essential to the improvement of your agriculture—as good transportation is essential—and that they ought be licensed for export as all other agriculture implements have been granted export licenses. I have good hopes and good confidence in the appeal I am making in your behalf in this item. Actually, the same situation exists in Hungary, Rumania, and Bulgaria; they need far better farm roads, far better transportation of agricultural products. . . .

Very sincerely yours,
Roswell Garst

May 5, 1961

To Chairman Nikita Khrushchev,
The Kremlin
Moscow, USSR

Dear Chairman Khrushchev,

We have just had the pleasure—we are, in fact, now enjoying the pleasure—of five people from your country, that is, Mr. Emelianov and Mr. Peter I. Pogodin from your Washington Embassy staff and Mr. P. G. Shmakov, president of Amtorg; Mr. Danilov, an economist . . . and Mr. Shpota, a chemist. . . .

They arrived by air yesterday at Des Moines and I personally met them and spent the afternoon showing them the newest in farm equipment—two new eight-row corn planters equipped with all the modern devices so that we can plant the corn—fertilize it—and put on herbicide and insecticide all in one operation. One man with such a corn planter is able to plant 120 acres a day—nearly 50 hectares a day.

I showed them all of the experimentation we are carrying on for the coming year. Then we all went down to my home in the late afternoon to meet Mrs. Garst and my two sons and my nephew, Mr. John Chrystal, and had a very enjoyable time.

This morning we all had breakfast together and they have all left with Mr. Grettenberg, with whom all of your delegations have become so well acquainted because he has supplied all of the implements which have been purchased by your Department of Agriculture for experimenting in the Soviet Union. In fact, Mr. Grettenberg is with them now and showing them some of the most modern American road grading equipment which we use for our secondary roads. We had a good deal of time for visiting yesterday evening—and we did visit at length—very, very broadly. I only wish you could have been part of the group because our visits had to do with progress in agriculture for the Soviet Union.

I told the visiting delegation that the biggest single advance in American agrculture was the use by hybrid seed corn—which was the first of the major tools to be developed. Your country, of course, has taken that first major step—you, too, have hybrid seed corn and in quantity sufficient for your needs.

The second major step forward in the United States was very greatly increased mechanization of agriculture. Your country is now in a position to make very rapid progress in the mechanization of your own agriculture. May I express the opinion, however, that you have made much more rapid progress in large field machinery than you have in electrification of the small tools around the collectivist and state farms—automatic waterers, feed wagons, etc. However, I do know you are making rapid progress in all of these lines and that the amount of labor required per unit of agricultural production will rapidly be lowered.

The third big step forward in American agriculture was a gigantic expansion in chemical fertilizers which has been followed by major expansion in other chemicals useful in agriculture—insecticides and herbicides.

Actually, my sons and nephew, Mr. Chrystal, and I are inclined to think that rapid expansion of the whole agricultural chemical production capacity in the Soviet Union will give you just as big if not a bigger lift in production than the first two items—than the use of hybrid corn—and the increased mechanization.

You have one additional step that it seems to us all your country must take to improve your agriculture—and that is the improvement of your secondary road system—the country roads—the roads that lead to the small villages—the state-owned farms—and the collectivist farms.

We had relatively few roads in the country that were creditable until the last fifteen years. The result was that it was difficult to either haul the products of our farms to the railroad or to haul the necessary supplies of production from the railroads to the farms.

And so about fifteen years ago, the national government, the state government, and the county governments and the township governments all started concentrating on better and better roads, and as a result we can now drive almost anywhere any time of the year with safety and convenience.

The manufacturers of road-building equipment have made marvelous progress so that a very few men can now put up a mile of road a day, or therabouts—and when you spread that over a few years' time, you get practically every road in fine shape.

I called your attention to our secondary roads when you visited here in the fall of 1959—we could drive anywhere on roads that had been raised up above the level of the adjoining farmland and graveled; they are raised because we want the snow to blow off them in the wintertime, they are gravelled so that we have a surface we can drive on after a rain.

As your agricultural production goes up, you will need more roads for the delivery of your produce to shipping points. As your fertilizer use goes up, you simply must have better roads for the delivering of fertilizer to the farms.

I assured Mr. Emelianov and Mr. Shmakov that I would continue to be just as helpful as possible to them in obtaining the finest information possible both as to chemicals and as to other implements such as roadbuilding equipment which I think can do so much for your nation, in the agricultural field with which I am so familiar.

I was proud to have Mr. Emelianov present me with a copy of a book about the farm of Roswell Garst.

I deeply regret the continuing tensions between our two countries and I am hopeful that these tensions will be lessened to a point where they completely disappear. The whole world needs the benefit of the lessons that both countries can teach them about producing more food, more industrial goods: how to raise their standard of living.

If the Soviet Union, Western Europe, the United States and Canada, and a very few other more scattered parts of the world, could join in being helpful to the great parts of the world which need help, what a wonderful world we could live in.

If the Soviet Union, Western Europe, the United States, and the few other scattered industrial areas could join wholeheartedly, the standard of living of the whole world, including those countries themselves, could rise to undreamed of pleasures and usefulness.

I must repeat the sentence that I told you when Mrs. Garst and I visited with you in the spring of '59—I must repeat that sentence because it is my creed:

"For the world to continue spending something like 100 billion dollars a year preparing for a war that nobody wants, no one really expects, a war that no one could survive—that is global insanity. . . ."

Very sincerely yours,
Roswell Garst

July 6, 1961

To Mr. Boris Burkov,
Novosti Press Agency,
Moscow, USSR

Dear Mr. Burkov,

Thanks so much for the pictures and the copy of the article.

In only one place did you make a serious error: on the back side of page 4, where you quote me directly at the start of a paragraph as follows:

"The year we were married it took us no less than six minutes to *produce* a bushel of maize. . . ."

What I actually said was as follows:

"The year we were married it took us no less than six minutes to *pick* a bushel of maize. . . ."

It was an error undoubtedly caused by language difficulties. It actually used to take us nearly thirty minutes to produce a bushel of corn, which included the full year-around work. Now it takes us five minutes, which is less time than it used to take for harvesting after all of the previous work had been done.

On the whole, both Mrs. Garst and I thought the article was unfair in its appraisal of American agriculture. We feel sorry for you—because apparently you feel that you must detract from the accomplishments of any country that has made greater progress than the USSR in order to build up your own morale—and that of your citizens.

Actually, the bulk of our labor is not hired by the hour—it is hired on [an] annual basis. The workmen live in completely modern homes with electricity, running water, baths—every one of them has a car; every one of them has a television set; every one of them has a refrigerator. They are as well housed—and as well paid—as top industrial employees.

I think you might well have pointed out that Mrs. Garst does her own housework—that she did, on the day you were here, feed twenty people or thereabouts without any help at all—and on a two-hour notice.

Furthermore, your description of the empty houses on farms in the United States is one you had better not make fun of—because Chairman Khrushchev is most anxious to imitate the United States in the reduction of the labor cost of raising food and fiber. He is anxious to be able to raise more food with less labor, and that's all we are doing here in the United States, and that is exactly what you are striving to do in the Soviet Union.

Seventy-five years ago, it took more than half of the Americans to raise the food and fiber. Thirty years ago it took one-third of the Americans to raise the food and fiber. Now it takes 10 percent of the Americans to raise the food and fiber—*and to raise more than we can use!*

It currently takes about 50 percent of the population of the Soviet Union to produce the food and fiber. Everybody connected with the Soviet Union ought to hope that this requirement is reduced to 25 percent of your population as rapidly as possible. That means you will have empty houses in the rural areas, in the villages; it means that the population of the rural areas will go down and the population of the industrial areas will go up.

The people who left these homes did not become "tramps." I think you know that. They moved into industry in the cities. They are building more electric appliances, more schools, better homes, better roads, more hospitals, more air-conditioning systems.

This is the exact direction that the Soviet Union ought to go and will go.

For you to try to make a bad picture out of American agriculture is ridiculous—and not helpful. I frankly resent it. Furthermore, I think Chairman Khrushchev will resent it. He himself has said that what the Soviet Union needs is an Iowa Corn Belt—the production of more meat, milk and eggs, and chicken meat. He himself has said that you need to produce more food with less labor.

And that's what you will do some day—and not because of people like yourself—but in spite of people like yourself. That's why Mrs. Garst and I feel sorry for you this morning.

Would it not have been much better if you had written an article saying that in the United States one man can cultivate one-hundred hectares—and do it without hard physical work on his part because of mechanization? Would it not been better for you to say that Russia is gaining rapidly by mechanization and that in the not too distant future a farmer in the Soviet Union will be able to do this very same thing?

Would it not have been better for you to have told the people in the Soviet Union that we are only using 10 percent of our total population in agriculture whereas you are still using 40 or 50 percent of your total population in agriculture—and that if you were to have lots more consumer goods, you simply must learn how to raise more food and fiber with less labor so that a higher percentage of your total population can be making refrigerators and all of the things that you need?

Why should you not point out that efficiency in agriculture has made a major contribution to the generally high standard of living that does exist in the United States?

The Associated Press picked up your story and cabled it to the United States. The papers who use the Associated Press article will quite generally comment upon it as an example of propaganda—and rather stupid propaganda at that. . . .

Very sincerely yours,
Roswell Garst

cc: Harrison Salisbury, *The New York Times*

December 26, 1961

To Gennadi Yassiliev,
Tass News Agency
New York, N.Y.

Dear Sir:

Your telegram of December 22 reached me this morning. . . .

. . . Many people in the United States do not realize that as a great nation, you are not as fortunately located as the United States climatewise. Your nation, as an average, is a little too far north and is a little too dry in most of its areas so that I do not think your production per acre in most of the Soviet Union can be expected to compare with the yields in the United States. But you can and will make terrific progress in spite of the climate handicap.

We have a vast area in the central part of the United States called the Corn Belt which is quite like your Kuban area except for the fact that we probably have a little more moisture than the Kuban and a slightly longer growing season.

In 1933 we had one-third of our people living on farms. That has gone down now to approximately 10 percent living on farms. At the same time our population has gone down, our production has nearly doubled.

Presently you have probably more than 40 percent of your people on farms. This will undoubtedly go down rather rapidly in the 1960 to 1970 period. I will be surprised and disappointed if you do not cut the percentage of people engaged in agriculture by one-half between now and 1970.

In Chairman Khrushchev you have a man who realizes the possibilities of agricultural progress. That is extremely important for your country. He apparently is insistent upon even more rapid agricultural progress than you are currently enjoying. It is excellent to have the leading political figure of any country give so much of his time and his energy and his thoughts to the production of more and better food because food is, after all, one of the world's necessities. Good food, ample food, food that is available at an economical price—those things are very basic in a world whose population is expanding as rapidly as our world's population is expanding.

The world now contains something like three billion people. Something like one-third of them go to bed hungry every evening. It has been my hope that the agricultural progress being made in the United States and the Soviet Union can serve as an example to the rest of the world so that all people of the world can be fed better. Pretty much, I have been interested in only one thing and that is teaching people to produce more food and better food with less labor. I think the Soviet Union and the United States ought to be engaged in that field.

And so my hopes for 1962 are simple—that the Soviet Union and the United States learn how to produce more food with less labor, and that all the other people in the world learn how to produce more food with less labor.

The insanity of the world at the present time is the spending on this globe something like 100 billion dollars preparing for a war that nobody wants, nobody expects, and very few could survive. That is global insanity.

In 1962, may we get a reduction in the armaments burdens of the world so that we can get on with the prospects that bring us hope instead of hopelessness—and bring us a better living.

Very sincerely yours,
Roswell Garst

June 6, 1962

To Chairman Nikita Khrushchev
The Kremlin
Moscow, USSR

Dear Mr. Khrushchev,

The American press and radio have, for the past several months, been reporting that you are disappointed in the rate of progress with agricultural production not only in the Soviet Union but also in Bulgaria. I am sure that you have made progress and that you will continue to make progress, and I hope that you make even more rapid progress than you have made—and I feel sure you will.

Compared with many other countries, the progress made by Soviet agriculture is phenomenal. It took the United States approximately fifteen years—between 1930 and 1945—to go from practically no hybrids of corn to practically 100 percent use of hybrid seed. It took the Soviet Union only from 1956 to 1961. I have discovered, for instance, that in India the use of hybrid seed still accounts for less than one percent of their total corn acreage.

But I never did believe that it was a very fast job. I thought it would come—progress in Soviet agriculture—at a slower pace than you antici-pated, because I knew that it takes much greater capital investment than you had realized and that capital investments are slow.

Fertilizer production facilities are very large—they can't be built in a few months—it takes a year or two or even three. It takes several years to get improved farm machinery perfected and manufactured and into use. I admire your determination to increase the progress of Soviet agriculture—your determination to accelerate the rate of progress—and I have confi-dence that you will make much more rapid progress from now on than even the substantial progress you have been making.

Because of the statement you made in visiting with Mr. Gardner Cowles when he visited with you recently—the statement that you would like to come to Iowa disguised as Mr. Ivanovich, disguised in a beard and a mustache, so you would not be surrounded by photographers and re-porters—I have been thinking that perhaps I could be helpful to you in your thinking and to the people in the Ministry of Agriculture if I came over to the Soviet Union for a brief visit.

Because of the pressure of my own business interests here, I would like to come over at your convenience some time between August 8 and the first of September. And either on my way over or on my way back, I would like to stop in Bulgaria, Rumania, and Hungary—not for a long visit but for a short visit with the Ministry of Agriculture in each case.

There are quite a few new things coming up in American agriculture that could be most helpful to the Soviet agricultural situation. I would like to discuss them with you—and with the people in the Agricultural De-partment.

The most significant, I believe, is the fact that we can easily double cattle numbers without any great change in our cultivated acreage by using the offal of the corn plant, the grain sorghum plant, and even by using the offal of the small grains—wheat, barley, oats, and rye. Some new equip-ment is available that you certainly must have. We are getting more and better information from the agricultural colleges on how to properly supple-ment the waste parts of our grain crop so that they can be more properly used for increasing numbers of cattle. It's one of the things you need most to learn about.

It is simply an extension and an improvement in methods of using urea as the protein supplement of our cattle in combination with the waste products of our grain crop. We don't need to use urea with molasses alone—we can use urea in combination with any fast carbohydrate which makes it much simpler.

In Bulgaria—and Rumania—they need definitely to embark on increased irrigation and I presume this is true in quite a few parts of the Soviet Union where irrigation is practical. Irrigation without heavy fertilization, without the use of proper insecticides—irrigation without the added tools of higher production—is not more than 25 percent as effective as irrigation with the proper tools.

I was very pleased to note in the American newspapers the other day that you were stressing the need for better secondary roads. Mr. John Chrystal—my nephew, who spent sixty days in the Soviet Union and Rumania and Hungary and Bulgaria in 1960—and I spend many hours discussing which things are needed most and soonest, and I put, as he does, improved country roads high on the list of necessities for agricultural progress. . . .

From all of the things I can read, you are now realizing to a greater extent than heretofore the necessity for greater capital investments in agriculture to speed up the rate of progress of Soviet agriculture. . . .

Very sincerely yours,
Roswell Garst

November 30, 1962

From Anastas Mikoyan[1]
Deputy Premier
Washington, D.C.

Dear Mr. Garst:

Availing myself of my short visit to your country I want to send you greetings and best wishes from N. S. Khrushchev, to which I add my own.

---

1. Mikoyan had just come from Havana, where he conferred with Fidel Castro about the missile crisis.

We all value highly your efforts at cooperation between our countries in the field of agriculture. Being optimists by nature we hope that after the most serious crisis in international life that took place is fully overcome,[2] new opportunities will be opened for expanding cooperation, among other things, in the field of your interest.

I was told that you were ill and I want to wish you sincerely full recovery, strength and cheerfulness.[3]

Thank you for remembering me. I was very pleased to receive your greetings.

Sincerely yours,
Anastas Mikoyan

MARCH 28, 1963

CABLEGRAM TO CHAIRMAN NIKITA KHRUSHCHEV
MOSCOW

THANKS FOR YOUR WIRE OF GOOD WISHES WHICH REACHED ME WHILE MRS. GARST AND I WERE ON VACATION. MY GENERAL HEALTH IS EXCELLENT, AS GOOD AS EVER, BUT MY VOICE WILL NEVER BE AS GOOD AS FORMERLY, ALTHOUGH I CAN SPEAK UNDERSTANDABLY.[1]

MY NEPHEW, JOHN CHRYSTAL, WHO SPENT SIXTY DAYS IN THE SOVIET UNION IN 1960 AND WHO ENTERTAINED GITALOV AND PARTY HERE IN THE SUMMER OF 1958, HAS RECENTLY BEEN APPOINTED TO THE IOWA BOARD OF REGENTS, WHICH BOARD IS IN CHARGE OF THE STATE UNIVERSITY SYSTEM.

HE AND I COULD TAKE TIME FOR A RATHER SHORT VISIT TO THE SOVIET UNION THIS SPRING IF YOU DESIRE AND IF YOU THINK WE COULD BE OF ANY ASSISTANCE TO YOU IN INCREASING AGRICULTURAL PRODUCTION.

PROBABLY THE LAST HALF OF MAY WOULD BE THE MOST CONVENIENT TIME FOR US IF THAT WOULD BE A CONVENIENT

---

2. The Cuban missile crisis of October 1962.

3. In the autumn of 1962, Garst underwent an operation for cancer of the throat that forced him to cancel his visit to the Soviet Union.

1. Garst was now using an artificial voice box, pressing the battery-powered device against his throat when speaking.

TIME FOR YOU AND THE PEOPLE YOU WISH US TO VISIT WITH IN
THE MINISTRY OF AGRICULTURE.
WE ARE BOTH SO BUSY THAT WE WOULD LIKE TO LIMIT
OUR VISIT TO NOT MORE THAN TWO WEEKS.
WITH APPRECIATION FOR YOUR WIRE OF GOOD WISHES I
SEND MY BEST REGARDS.

ROSWELL GARST

May 15, 1963

From Nina Khrushcheva
Moscow

To Elizabeth Garst
Thank you for the books and other things that I've got from your
husband in my house.
I send some foto-album of Moscow and Leningrad. I hope you had no
time to take such ones when you were here.
Best wishes to you and the whole family.
Excuse my bad English. I'm forgetting the language every month and
every day. 63=is the age![1]

Be happy!
Nina Khrushcheva

June 5, 1963

To Chairman Nikita Khrushchev
and Family
The Kremlin
Moscow, USSR

Dear Mr. Khrushchev,
Because Madame Khrushcheva and your daughters read and speak
English well, I am sending herewith several bulletins[1] which I believe will
be of highest interest to you and to them.

1. Holograph letter, brought back from Moscow by Garst after the 1963 visit.
1. The three bulletins summarized in the letter were all written by Garst for distribution to his customers and other interested parties.

I have sent similar bulletins to the Soviet Embassy in Washington for translation and asked them to forward the copies to [A. S.] Shevchenko, to N. K. Baibakov of the State Committee for Chemicals and Petroleum; to the Minister of Agriculture [Ivan P.] Volovchenko; to Governor Leonid Adrienko of Kiev; to Vasilii Khomiakov, chairman of the Bureau of Agriculture of the Ukraine; to Mark Spivak, minister of agriculture of the Ukraine; to Valery Tereshtenko; and to Mr. Dzivbanov of the Kiev Region.

It seems to me that every one of these bulletins is of the highest importance to your agriculture.

The bulletin *Congratulations* spells out the extremely rapid progress in production since we have had fertilizer and insecticides and herbicides in quantity. The figures of yields are the average yields for the United States. The U.S. average yield never exceeded 40 bushels per acre until 1958 and now in 1961 and 1962 it was above 60 bushels per acre. That terrific increase was mostly chemical!

The second bulletin *Progress and Opportunities in Corn Growing* is a record of the advantages of mechanization and the lowering of man-hours to produce a bushel of corn in the United States.

It takes about the same amount of time to farm an acre or a hectare whether the yield is high or low. So if you or we double the yield, we cut the time required per unit of corn or silage by half.

Because on the average we have better rainfall than your country, we will likely always have higher average yields and lower unit labor costs—but your rain should be as fine percentagewise as ours has been.

And finally, the third bulletin, on how to use urea and molasses as a protein for cattle and sheep and goats, is of extreme importance to your country.

We are fortunate in the United States in having a very large area where soybeans yield very well. As a result, we are now producing more than 500 million bushels per year. Beans do not yield well in dry areas—so they are not a good crop in most of your country. Sunflowers, which yield well for you, are high in labor cost. The result is that we have more protein of vegetable form available at the present than you. But even with rather ample supplies of soybean oil meal, we did use 150,000 tons of urea last year as the protein of our cattle and sheep. And even then we did not use as much as we should have used.

As pointed out in the bulletin, you can and should use urea for 100 percent of the supplemental protein. The corn silage—whether you put it up in the form of green mass or in the form of whole corn silage—carries from 5 percent to 8 percent protein. (Incidentally, if you apply a heavy nitrogen fertilizer to the corn field, the corn crop harvested for silage will have a good deal more protein than if not heavily fertilized).

Beef cattle should have 10 to 12 percent protein in their diet. Dairy cattle should have 14 to 16 percent in their diet. Urea can be used for all of the additional protein needed.

By doing this—by furnishing your cattle and sheep with all the supplemental protein in the form of urea—you could liberate the sunflower meal, flaxseed meal, and cottonseed meal that you have been feeding your cattle and transfer these by-product oil meals to your swine and poultry.

Mr. Chrystal and I enjoyed the progress in agriculture not only around Kiev but in Hungary and Rumania as well. We had a wonderful trip all the way.

We are mindful of the numerous things you inquired about such as a plant for the production of vitamin A, a blending plant for dry fertilizer, a plant making premix feeds, etc., etc., and are even now engaged in finding out about them. We will be in touch with Mr. Emelianov.

Please take this letter home to Mrs. Khrushchev, to whom we wish to express our deep appreciation for her hospitality, and to your whole family. Mrs. Garst was highly pleased with the pictures of Moscow and Leningrad and asked me to send her thanks to Mrs. Khrushchev.

Respectfully yours,
Roswell Garst

August 13, 1963

To Chairman Nikita Khrushchev
Moscow

Dear Chairman Khrushchev,

Both Mr. John Chrystal and I want you to know how pleased we are about the atomic test ban agreement. It appears certain that it will be confirmed by the Senate and rather soon.

Also, we wanted you to know that Mr. and Mrs. S. S. Malov, president of Amtorg, and Mr. and Mrs. Emelianov drove out for a vist about ten days ago. We all had a fine visit and we were able to show them another very fine crop of corn and sorghum, our cattle feeding operation, our hogs, and a very great deal of Corn Belt agriculture.

We have already submitted full information on equipment for the building of what we in the United States call "farm to market roads." You doubted that we could get export licenses for such road-building equipment, but both Mr. Chrystal and I are confident we can get the export licences, and I urge you to have the order placed at once.

We also gave Mr. Malov and Mr. Emelianov prices and specifications on a pre-mix feed plant and an alfalfa pelleting plant—and of the two-row self-propelled silage harvesting equipment such as Mr. Shevchenko drove when he was here last year at this time.

We have made preliminary investigation about a plant for the manufacture of vitamin A and also for an ammonium phosphate plant and should be able to submit detailed information shortly.

I am having one of our very best men meet your sorghum delegation in Oklahoma and Kansas. He will show them the best cultural practices and show them irrigation of both corn and sorghum. We expect to have them visit here in Coon Rapids as well.

I do hope you will encourage the immediate placing of the order for the road grading machinery because I want to get that export license to prove to you that the United States will issue it.

Because Mr. Shevchenko is very familiar with this whole Corn Belt type of agriculture, I wish to suggest you send him over to conclude the purchase of the numerous items we discussed with you in May.

The ban on atom bomb testing is a great step forward in the relations between our two countries. I trust it is, however, only the first step.

Very sincerely yours,
Roswell Garst

December 2, 1963

To Nikita Khrushchev
Moscow

Dear Chairman Khrushchev,

. . . I am terribly sorry that the widespread drought so seriously reduced the yields to the point where you have had to spend more than $500 million for wheat and other grains.

With the fertilizer plants you are building and the ones you will build, all you are going to need is normal rainfall to build up reserves against the occasional dry seasons which will come.

I have no doubt at all about the future productiveness of Soviet agriculture. We in the United States were short of food as late as 1951. We did in the 1950–1958 period what you are doing now—we built fertilizer plants with great rapidity and roads for the easy distribution of the fertilizer—and as a result we are now producing surplus foods. I predict that will happen in the Soviet Union as it did happen in the United States. . . .

The Soviet Union is going to spend this year something above $500 million for agricultural supplies.

I sincerely believe that if you had spent 2 percent of that amount, $10 million per year for the last eight years and built the fertilizer plants sooner, you might have saved a substantial part of this year's expenditure.

As you know, I have urged you each time I have seen you to increase your purchases of the newest and best tools of agricultural progress which we have available. . . .

Everyone in the United States was stunned and saddened at the assassination of president Kennedy. He was so young and fine that even those who disagreed with him politically were as shocked and grieved at his death as were his admirers. It was the work of a madman.

The sympathy expressed by you and Madame Khrushcheva and the Russian people was highly appreciated by everyone in the United States.

President Johnson is as interested in peace as was President Kennedy, in my judgment. I look forward with confidence and predict that relations between our two countries will continue to improve. . . .

Sincerely,
Roswell Garst

[January, 1964]
From A. S. Shevchenko
Moscow

Dear Mr. Garst:

Nikita Sergeevich Khrushchev received your letter of December 17, 1963 and with an interest familiarized himself with the ideas expressed by you. May be Nikita Sergeevich himself will write you a letter, but now he is very busy with urgent affairs. . . .

In the connection with your letter I would like to make my personal comments. You consider it desirable that Soviet agencies would buy in the U.S. road construction machinery for building as you say "from farm to market" roads.

Road construction equipment in the U.S. is really good. When we were on your farm we saw it in operation and were satisfied [a] great deal. As you know our country is on such a level of industrial development that it is producing such machinery. And, as experts told me, the road construction machinery made in the USSR is of high quality. But now it is a matter of investment rather than technical abilities. It seems to me that the reason why in some areas of the USSR there are no good roads is not that we produce few road construction equipment, but because we [are] still short of investment for construction of roads.

Now, as you know, our government focuses its attention on sharp increase[s] in the production of fertilizers, herbicides as well as on irrigation projects. For this huge investments are being earmarked which, of course, does not permit to find required resources for simultaneous development of road construction on a large scale.

The program of the chemistry development in our country set forth in N. S. Khrushchev's report makes all of us agriculturist[s] happy and opens up good prospects in future. This is why buying chemical equipment is now a matter of great interest for us. As I know, Soviet foreign trade agencies have received many offers from different firms in capitalist countries to supply chemical equipment. These offers are being studied.

I don't know whether our agencies would buy equipment for chemical plants in the U.S.A. There are many obstacles in that country. Purchase of equipment is connected on one hand with reasonable prices, which at any rate should not be higher than prices on the world market, and on the other hand granting a loan. In this instance American legislation creates difficulties.

Of course, if the barriers created by the Congress in the way of the Soviet-American trade are removed, and if American prices of equipment are acceptable, our foreign trade agencies apparently will be able to buy some chemical equipment in the U.S.A.

But if the barriers are not removed, firms from other capitalist countires will sell such equipment to us at existing prices or may be even cheaper and on better terms. As I have already told, Soviet foreign trade agencies have many suggestions from different firms.

You write that there is opportunity to buy in the U.S.A. a plant for production of pre-mix feeds, an installation for vitamin "A" synthesis, several sets of best machines for distribution of fertilizers and machines for crop production. These suggestions are very interesting ones. I believe that appropriate Soviet agencies would be interested in them.

The following words in your letter to N. S. Khrushchev made me very happy.

"We would like to be helpful to you and Soviet agriculture." It is very good. N. S. Khrushchev says that mutual exchange of experience enriches the peoples and helps to move further successfully. . . .

I have a request to you.

In the middle of February the Plenary meeting of the Central Committee of C.P.S.U. will take place. During this meeting the problems of intensification of agriculture through wide application of fertilizers and mechanization will be discussed. In this country you are known as an outstanding organizer of highly intensified and efficient farming operation.

We would appreciate very much if you could write for Soviet press, as soon as possible, a big detailed article about your experience in farming, about your farm and general trend in farm management in the U.S. from the standpoint of effective use of land, feeds and labour.[1]

I name, of course, this topic very relatively. You are a man of great knowledge, big practical experience and, no doubt, you will find what to say to our agricultural people. I am sure that your article would have been met by many readers in the Soviet Union with great appreciation.

Dear Mr. Garst, it is a great pleasure for me on behalf of N. S. Khrushchev to convey to you and your family his hearty wishes. Comrade Khrushchev often remembers meetings with you, interesting talks, your fine farm and your warm hospitality during his stay in Coon Rapids. . . .

Sincerely yours,

A. Shevchenko

January 28, 1964

From A. S. Shevchenko
Moscow

Dear Mr. Garst:

Lately I sent you a letter which I believe you already received. I mentioned there that our farm specialists who met you always with great warmth remember you and advices you gave to them. And quite recently I was present at one talk during which once more the job being done on your farm was appreciated. I would like to share some ideas with you.

During the meeting of Nikita Sergeevich Khrushchev with Fidel Castro in Moscow the talk turned on the development of agriculture on Cuba and use by Cubans of experience in farming of other countries.

1. See Appendix.

In this connection Nikita Sergeevich Khrushchev for a long time and in detail told Fidel Castro about his meetings with you, about his trip to Coon Rapids, and your fine family. He told much about your management of the farm, how you grew hybrid corn seeds and produce beef. The talk also turned to mechanization of crops and livestock on your farm. Nikita Sergeevich Khrushchev especially emphasized your ideas on soil fertilization, the role of chemistry in agriculture, continuous corn production etc.

Fidel Castro showed a keen interest in all that and said that he reads American farm publications and finds many interesting things in them.

It seems to me that Fidel Castro revealed a big desire to familiarize himself closer with your experience. I don't know whether it is timely but I thought wouldn't you find it expedient and possible to make a trip to Cuba and familiarize yourself with agriculture there and its possibilities. I think it may turn to be a very interesting trip. I don't know how to do it and whether it is possible or not. Apparently nowadays it is difficult to do. Of course, it is easier for you to see the whole picture.

Now we think of the next year to come. It is difficult to make forecasts now but apparently moisture supply will be better than last year. In the fall we had much more rain. Now we have real Russian winter with good frosts. Though some time there are considerable fluctuations; 5 degrees below the zero in day time and 20 degrees at night. In the extreme North there is as low temperature as 60 degrees C. below the zero.

It is interesting to know what weather you have in Iowa? What results have you got in the end of the year? What is about your health?

Please, convey my best wishes to your wife, your sons—Steven and David, and your nephew—John Chrystal.

Sincerely yours
A. S. Shevchenko

February 4, 1964

To Mr. A. S. Shevchenko
Moscow

Dear Shevchenko,
Thanks so much for your letter of January 28 telling me that Chairman Khrushchev had told Fidel Castro all about me and that Castro would like to know more about the new things in American agriculture.

As you know, I have always felt that plentiful food—a full stomach—

makes the largest contribution toward bringing peace to the world and that by using modern agricultural techniques, there is "no need for hunger."

I have just forwarded you a book—written by my brother, Jonathan, with that title—*No Need for Hunger.*[1] I sent it through the Soviet Embassy in Washington and it should reach you soon. I also sent copies to Madame Khrushcheva, Marina Rytova, and Tereshtenko.

Because I sincerely believe that there is no need for hunger and that my very large and long experience gives me a wide background that helps me teach others how to use new techniques, I have been spending a good deal of time at it.

While I am a capitalist—not a Communist—I have felt that hunger itself affects people regardless of race, creed or color—or political affiliation. And the fight against hunger should be carried on everywhere. I have always said I want to help people "learn to produce more and better food with less labor" and that I would do so regardless of race, creed, color, or political affiliation.

However—

As an indiviual citizen, I have to pay attention to the attitude of my government and my fellow countrymen.

As long as Castro accuses the United States of aggression—and all manner of other unscrupulous activities—as long as he promotes anti-American activities in the other countries of Central and South America, no American citizen will offer to be helpful to him—nor should do so.

Castro is not trying to bring peace to the world—he is trying to promote revolutions that will actually bring more hunger, more tragedy.

I would be perfectly willing to help the Cubans and Castro if he indicated that he wanted peaceful coexistence, as has been expressed by Chairman Khrushchev.

But, until he changes his attitude, I think no American will be willing to be helpful. . . .

Sincerely
Roswell Garst

1. New York: Random House, 1963.

March 27, 1964

From Ilya Emelianov
Agricultural Counselor
Soviet Embassy
Washington, D.C.

Dear Mr. Garst:

. . . Quite recently I've got "Izvestia"—paper dated March 7, 1964 where the Report of the Head of the Soviet Government Nikita S. Khrushchev on intensification of agricultural production is published. There in several places Nikita S. Khrushchev referred to the experience of the U.S.A. including your experience on fertilizers (1), corn breeding by one person on 100 hectares, study of your experience by our specialists—Gitalov and others (3). By the way, in 1963 in Krasnodar region the team headed by V. Pervitsky cultivated corn according to your method on the square of 445 hectares. In average, one person cultivated corn on the square of 148 hectares. Average yield was equal to 51.5 centner per one hectare (72.6 bushels per acre). Labor expenditures per one centner (about per 4 bushels) of grain accounted for 8 minutes or about 2 minutes per one bushel.

In two other instances Nikita S. Khrushchev referred to the United States experience on organization of the extension service (2), broiler industry (4–5) (commercial Farm—La Torre, California, near San Francisco).

I marked with a red pencil the corresponding places in the Report of Nikita S. Khrushchev & indicated the same numbers in this letter in brackets. I am sending a copy of this paper to you. The Report of Nikita S. Khrushchev was also published in "Pravda", "Selskaya zhizn" and in the others.[1]

I wish you a happy journey, good luck in your trip and happy return.

Please convey my and my wife's warm regards and best wishes to the families of your sons, to John Chrystal and Mr. Grettenberg with his family.

Sincerely yours,
Ilya Emelianov

1. In the margin of the *Izvestiia* article, Emelianov apparently numbered from one through five Khrushchev's references to the U.S.A. In this letter, he cites numbers one and three first because they refer specifically to Garst; he then cites two, four, and five, which refer indirectly to Garst and his advice. The article is missing from the correspondence.

April 8, 1964

To Mr. Philip Maguire
Washington, D.C.

Dear Phil,

. . . Because I am leaving for a couple of months in Central and South America helping the AID part of the U.S. State Department teach the people of South and Central America how to raise more food with less labor, I want to review what you and John Chrystal and I did a year ago.

Ever since 1955—my first trip—I have tried hard to get the building of what we in the United States call "farm to market" roads started in the Soviet Union.

They have a few highways that are fine—but the bulk of the country roads are just as bad as, or worse than, the roads we had in the Midwest fifty years ago.

Because I did not want to encourage Chairman Khrushchev to order American road building equipment unless an export license would follow, we talked to Ambassador Llewellyn Thompson and to the people in the Department of Commerce, including Secretary [Luther] Hodges himself.

Furthermore, you will remember that I had planned to go in the fall of 1962—but could not go because I had my larynx removed—but in the fall of 1962 we did talk to Ambassador Foy Kohler while he was still in Washington.

Now finally, eight years after I started talking, the Soviets have asked for quotations on three sets of American road building equipment—each set large enough to construct about a mile of farm to market road per day, when building roads through either level or only gently rolling areas. I have assured them that we can send an experienced operator along to train them in the use of such machinery—as we did with the Rumanians when we sent the farm machinery over in 1956.

Because we did talk to both Ambassadors Thompson and Kohler and to Secretary Hodges, I believe there should be no difficulty, but I want no hitch nor trouble, and I think it may be well for you to remember our care in seeing that the matter was discussed in advance. . . .

There is a great deal of satisfaction in seeing that Chairman Khrushchev is talking about more "consumer's goods," more about "better living conditions," etc. etc. The more he talks that way, the more difficult it will be for him ever to talk any other way.

He is whetting the appetites of every person in the Soviet Union and in the other Communist countries of Eastern Europe for better living. That appetite will have to be satisfied.

As you know, and as both Ambassadors Thompson and Kohler know, this kind of progress is what I have been trying to promote.

Good roads are a part of the picture. A few good roads call for a lot more good roads and finally for good roads everywhere.

As it was in the United States, so it will be in the Soviet Union. The demand for good roads is hard to satisfy!

That's why there must be no hitch in the export license.

Sincerely,

Roswell Garst

December 12, 1964

To Mr. and Mrs. Nikita Khrushchev
Moscow

Dear Mr. and Mrs. Khrushchev,

Both Mrs. Garst and I wish to extend greetings to you both as the new year of 1965 approaches.

But even more, we wish to tell you what a great contribution you did make to better understanding between the USSR and USA during the time you were chairman.

Before you, Mr. Khrushchev, became chairman, there had, for nearly ten years, been little or no contact between the people of our two countries. Ignorance of each other led to suspicions of each other.

The government of the Soviet Union put out propaganda to the effect that the American government was controlled by heartless capitalists who kept the bulk of the American people in poverty, which was, of course, not true.

The American politicians pictured the people of Communist counties as the "slaves of Communism" to quote a phrase from Vice President Nixon while he was visiting the Soviet Union.

It was your talk before your Congress in February of 1955 in which you said that what the Soviet Union needed was "an Iowa Corn Belt" that broke the ice.

The editor of the *Des Moines Register*, Mr. Lauren Soth,[1] wrote an editorial saying that if you needed an Iowa Corn Belt, you should send a delegation over to find out how an Iowa Corn Belt is operated.

1. Soth was editor of the editorial page of the *Register*.

The result was the first exchange of farm delegations in 1955. That was the forerunner of many many exchanges that followed. Thousands of Americans have since visited the Soviet Union and thousands of people from the Soviet Union have visited the United States.

Unfounded suspicions such as those that did exist before 1955 have largely disappeared.

You both—and most people in the Soviet Union—now recognize that our form of government works very well for us, that our agriculture and industry have given all Americans a high standard of living, that our democracy has no desire for war or for conquest.

We, on the other hand, recognize that your country is also highly interested in progress, in getting higher living standards for all of your people; that the Soviet Union suffered so terribly in both World War I and World War II that you have every reason not to want war, and a sound reason for being sure that if war ever does come, you will never again be overrun.

In short, those years during which you were chairman saw a very great lessening in the possibility of conflict of arms between our two countries— because we know each other better.

So—

Mrs. Garst and I think you have made a great contribution to the peace of the world . . . .

Sincerely,
Roswell Garst

December 12, 1964

To President Anastas Mikoyan
Moscow

Dear President Mikoyan,

. . . I do want you to know that I am just as willing to be helpful to the agriculture of the Soviet Union since the retirement of Chairman Khrushchev as I was before. I have so told Ambassador [Anatoly] Dobrynin and Mr. Emelianov when I was in Washington early this week. I am so telling you via this letter.

I will appreciate knowing from you if there is any way I can be helpful to you or to your minister of agriculture. . . .

. . . Your country and my country—our two countries—ought to be friendly countries.

The situation is greatly improved when compared with the fall of 1955 when I first met you and Mr. Khrushchev.

But further improvement is most desirable.

I want you to know that I hope 1965 will see continued improvement in the relationship between our two countries—and that I will do what I can to help. . . .

Sincerely,
Roswell Garst

USSR, 1969–77

In 1969, a new period in the Coon Rapids–Moscow connection was heralded by the visit to Iowa of economist Marina Menshikova. In 1971, the year of Khrushchev's death, Vladimir Matskevich made the journey to the United States long sought by Garst and returned to Moscow with samples of hybrid sorghum seed. In 1972, a purchasing delegation arrived in Coon Rapids, shortly thereafter completing negotiations for a substantial purchase of seed. Then followed negotiations for sorghum processing plants. Garst and Chrystal were invited to the USSR in 1972 and again in 1974. Delegations continued to arrive through 1977.

The renewed Soviet interest in Iowa apparently reflected a Politburo decision in 1971 to pursue with renewed vigor, and on a much larger scale, the acquisition of Western technology in all fields. The correspondence of Dmitriy Polyansky, the new minister of agriculture, makes this intention clear with respect to agriculture, as do the increasing purchasing authority and commercial sophistication of the visiting delegations. In the case of the sorghum plants, for example, it was evident that the Soviets were shopping in other Western countries as well.

Garst's correspondence is filled once again with technical details as well as broad recommendations. His visits to the USSR revealed that his professional stature as agricultural adviser remained undiminished by the departure of Khrushchev. His friendships remained firm, as the letters from Ivan Khoroshilov and Alexander Tulupnikov demonstrate. In the summer of 1977, Andrei Tulupnikov, the son of the man who proudly claimed that he had "discovered" Garst for Russia, went to Coon Rapids and received the same tour his father had been given twenty-two years before. He was the last Soviet visitor Garst entertained. Garst died on 5 November 1977, but the connection between Coon Rapids, Iowa, and the

Soviet Union continues to flourish, carried on primarily by John Chrystal. Chrystal still travels regularly to the USSR and recives Russian visitors in Iowa.

April 28, 1969

To Professor A. I. Tulupnikov
Director, All-Union Institute of Scientific
and Technical Information for Agriculture
Moscow

Dear Professor Tulupnikov:

We have been enjoying the visit of Prof. Mrs. Marina Menshikova greatly. She arrived in Des Moines Saturday evening. Mrs. Garst and I met here there and brought her to Coon Rapids Saturday evening. So her first meal with us was a Sunday morning breakfast. It reminded me of your being in our home for Sunday morning breakfast in 1955. You were the first man from the Soviet Union we ever met!

We spent all day yesterday discussing new things in agriculture—the progress that has been made and the progress that still should be made.

We showed Mrs. Menshikova our very extensive beef cattle operations. We are sure that by using the urea form of protein, along with relatively small amounts of some fast carbohydrate such as the sugar in molasses or the starch in grain, all common celluloses become potential feed for ruminants, the cattle and sheep.

This morning, Mr. John Chrystal, my nephew, who is president of the bank here at Coon Rapids and also the state superintendent of banks for the whole state of Iowa, is discussing agricultural credit with Mrs. Menshikova, and then I am taking her down to Des Moines to visit with the research departments of the Pioneer Hi-Bred Corn Company, about new genetic developments in corn, grain sorghums, forage sorghums, chickens (both laying chickens and broilers), and hybrid wheat.

Then I will take her up to Iowa State University at Ames.

Mrs. Menshikova and I are in agreement about a good many things— but especially about the fact that there should be more visits back and forth between our two countries.

Neither you nor Minister of Agriculture Matskevich has been back to the USA since 1955. It has been six years since I was last in the Soviet Union.

Mrs. Menshikova will not return to the Soviet Union for almost two months. She asked me to send you the folders that I am enclosing. . . . With warm regards to you.

Sincerely,
Roswell Garst

July 3, 1969
To Professor A. I. Tulupnikov
Director, All-Union Institute of Scientific
and Technical Information for Agriculture
Moscow

Dear Professor Tulupnikov:
. . . It seems to me that both the USSR and USA have much to gain from more visiting back and forth. For instance:

You and Minister of Agriculture Matskevich visited the USA in 1955. That is fourteen years ago—and I believe neither of you has been in the USA since.

In 1955, the USA was only starting to use fertilizer—especially nitrogen fertilizer on corn. A history of corn yields for many years tells the story rather well.

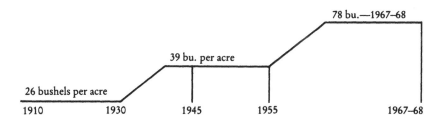

From 1910 to 1930 we averaged twenty-six bushels per acre. It took fifteen years—from 1930 to 1945—to change from varieties of corn to hybrids of corn. By 1945 we were using about 100 percent hybrid seed, and our yields had gone up 50 percent.

Then, we increased the production of nitrogen, phosphorus, and potash and improved and expanded and cheapened the price of fertilizer.

For the past two years the yield has been seventy-eight bushels per acre on the average for the USA. We now use all hybrid seed, generous fertilizer applications, chemical weed and insect control, ever-expanding irrigation—all of the most modern inputs.

Furthermore, we are using both grain sorghum and forage sorghum in our areas of low rainfall, and we are improving our knowledge of how to better use them all.

On the other hand, the USA is only now starting to take advantage of artificial insemination of cattle, which your country has used so widely for many years. . . .

Sincerely,
Roswell Garst

January 2, 1970

To Dr. Marina Menshikova
Institute of World Economics
Moscow

Dear Dr. Menshikova

Thanks for your card of seasons.

As I have recently written Minister of Agriculture Matskevich and Mr. Tulupnikov, it seems to me that the decade of the 1960s was a bad decade for both the people of the world—and the governments of the world.

And almost the whole blame can be charged to the excessive armament burdens.

Because I feel that both the people and the governments of the world are realizing that there is greater danger in an armaments race than in what might be termed a "disarmament race," I have reasonable hope that the decade of the 1970s may well see the greatest progress of any decade in human history.

The recent meetings between our two governments in Helsinki gave me added hope—until the final report, which greatly disappointed me. The final report said that the real meaningful negotiations would be resumed in mid April. Just why they had to be delayed four months, I fail to understand.

I was hopeful that they would start in mid January. But a start has been made—and I look foward with enough hope to start planning. . . .

I only know one subject very well—how to produce *more* and *better* food with less labor.

When you were here I suggested that the USSR needed several things in agriculture. More and better farm to market roads was one of these things. More fertilizer—and better distribution methods, and more irrigation, and better and more warehouses. And machinery.

In the USA we now have only about 7 percent of our total population on farms. We take a census every ten years—and it is probable that the census may show only 6 percent on farms. And we enjoy an excellent diet and export large amounts of wheat, feed grains, and soybeans.

In my recent letters to Mr. Matskevich and Mr. Tulupnikov, I urged both to come back to the United States themselves. It will have been fifteen years since they were here in 1955.

It seems to me they can send technical delegations to spend more time *after* they have been here. However, I wish to make a suggestion to them through you, and that is that I have practically retired from business—and turned it over to my two sons, Stephen and David, and to Mr. Chrystal. So, if either Mr. Matskevich and/or Mr. Tulupnikov does decide to come to the U.S.A. in 1970, as I believe they should do, I would be pleased to plan at least a major part of their trip—and to act as their guide. . . .

> Very sincerely,
> Roswell Garst

July 27, 1971

To Mr. William P. Rogers
Secretary of State
Department of State
Washington, D.C.

Dear Secretary Rogers:

. . . It seems to me that President Nixon's effort to get China to lessen its isolation is a grand step in the right direction. But—

I fear that unless we continue to cultivate our relations with Russia, we may be making a mistake.

It is in agriculture that the Soviet Union has perhaps its greatest opportunity for rather rapid progress.

Matskevich is a very keen and influential person. Why would it not be an excellent idea to have Secretary of Agriculture [Clifford] Hardin invite Minister of Agriculture Matskevich to the USA to see the progress in American agriculture since his visit here in 1955 with the first delegation? It seems to me that it would show our interest in the USSR—and in the whole world food situation. The United Nations recently forecast a doubling of world population between now and the year 2000. So better agriculture will be necessary everywhere.

It may be that inviting Matskevich over this fall might show that our interest in China has not lessened our interest in the Soviet Union.

Like ping-pong, which was used by the Chinese as a method of opening up signs of friendliness, agriculture was an excellent way in 1955—and can be an excellent tool in 1971.

If you think well of the idea, feel free to tell Secretary Hardin that I will be happy to be helpful to him in suggesting things I feel sure Minister Mastskevich would be interested in seeing.

The Soviet Union is already well advanced and very good at chickens, both broilers and egg layers, and at large scale swine production.

They need to see irrigation as we use it in the USA. They need to see large scale cattle feeding operations as we do it in the high plains areas.

And most of all, they need a transportation system. They need—and must have —a much improved farm to market road system—to transport tractor fuel and fertilizer to their farms and grain and livestock and poultry products to their population centers.

It might be well to invite not only Matskevich, minister of agriculture, but the man who heads up their Highway Department.

Matskevich is an extremely competent individual. He would be an influential person in the cabinet of any country. . . .

Very respectfully yours,
Roswell Garst

September 15, 1971

To Mrs. Nina Khrushcheva
and Family:

We send our sympathy to you all.

What a great life Nikita Khrushchev lived! What great contributions he made—not only to the Soviet Union but to the world.

Born under an extremely despotic czarist government that was afraid to educate the citizens for fear they would revolt, he died under a Communist government where universal education is required, and where literacy is universal.

Endowed with a strong body and an active mind and with great energy, he became a leader in the drive to make the USSR a great nation.

He undoubtedly made mistakes along the way—everyone who tries hard for progress makes mistakes. Between 1917 and 1964, the USSR changed from a very primitive, backward nation to one of the truly great nations of the world.

And Nikita Khrushchev was one of the principal causes of the change.

Not only did the USSR benefit from his life—so did the world. He established communications with the rest of the world. Good communications are important between nations—as well as between people.

We enjoyed our several contacts with him. Better than most men of prominence he knew how to laugh, which is a great virtue in this troubled world.

We send to all of you not only our sympathy—but our warm regards,

Very sincerely yours,
Roswell Garst
Elizabeth Garst
John Chrystal

October 14, 1971

From Nina Khrushcheva
1221002, Moscow
Starokoniushenny 19, kw. 57.

Dear Mr. GARST!
Thank YOU and YOUR family for
Your letter of condolences and sympathy.
We were glad to receive Your friendly
appreciation of my husband's character
and of his role in the world politics.
You wrote, that "N.S. knew how to laugh."
Yes, HE knew. It was his human trait,
which attracted people and facilitated
personal contacts with them.

N.S. highly respected You and valued
Your energy and skill in Your business.
                              With warm regards to You
                              and Your family
                              Nina Khrushcheva
    P.S. Please, excuse my bad English.
    Let me know, when You get my letter.

December 6, 1971

To Mr. V. V. Matskevich
Minister of Agriculture of the USSR
The Soviet Embassy
Washington, D.C.

Dear Minister Matskevich:

First, I wish to welcome you back to the United States. It has been a
little more than sixteen years since you visited here in 1955—too long.

Your visit is so short that it seems to me that the thoughts I have
should be submitted in writing so they can be interpreted into Russian—
and you can study them when you have more time—and also so you can
have your associates study them in more detail. Because the suggestions
cover different fields, I submit them to the several departments. And in
each case, I submit them in triplicate so you can keep a copy for your own—
and send a copy to your assistants.

For your information, I enclose the production of corn for grain in the
USA since 1950: the number of acres of corn for grain, the average yield in
bushels per acre, and the total production.

Please note that the year you brought the delegation here in 1955, we
produced only forty-two bushels per acre compared with eighty-six bushels
per acre this year and that our improvement has been rather steady throughout the whole twenty-two year period. I list the reasons for this doubling of
yields and give what I believe are something like the percentage increase
due to each item.

1. More fertilizer use by almost all farmers and better balanced
fertilizer—60 percent of the increase.
2. Better hybrid varieties—10 percent of the increase.
3. Better control of weeds and of insects through chemicals—
10 percent improvement.

4. Better machinery all the way through, which has permitted more timely planting, more timely harvesting—5 percent improvement.

5. Harvesting at somewhat higher moisture levels than twenty years ago—and drying for storage, that permits far less field loss and improved quality—5 percent improvement.

6. Irrigation. Because most of the USA corn crop is raised where precipitation is adequate for very high yields, irrigation does not affect—or greatly affect—more than 20 percent of the U.S. corn crop. But where irrigation is needed, it increases yields by four or five times the nonirrigated yields. Probably 10 percent of the increase has come from increasing irrigation of the last twenty years.

That is the reason I believe irrigation is so important; that is the reason I so strongly urge a series of delegations next summer. . . .

Very sincerely,
Roswell Garst

Idea No. 1

First, about your seed corn plant at Krasnodar:

In the winter of 1955–56, [to] the delegation that came over to inspect the Pioneer brand hybrid seed corn, which was purchased to give widespread background demonstrations of the superior yields and better strength of stalks of hybrid corns when compared with varieties, we also furnished blueprints of the hybrid seed corn plant that we had at Coon Rapids, and we helped the delegation purchase all of the machinery we used in our plant. . . .

In the intervening period, technical research has proved that the original fans we were using to blow the warmed air through the ear corn could be improved in efficiency by changing the vanes. We have changed all of our own. Then a further gain in efficiency can be obtained by doubling the horse power—which we have done. The result of these changes has been that we are able to gain substantially in the amount of corn we can dry through each bin. We think we are drying from 35 to 40 percent more seed corn through our bins than we were able to do before these rather small and rather inexpensive changes were made.

Also, we have better tools in our seed corn laboratories where we check the accurateness of our sizing of the several kernel sizes to permit accurate planting. . . .

In those days, we used far, far more labor than we use now. We use pallettes and fork lift trucks. We have mechanized almost as rapidly as labor has increased in cost per hour.

Garst & Thomas will be happy to be as helpful as possible in showing any delegation you care to send all of what we believe are improvements and labor saving to be of assistance in helping you order what new and more modern equipment you may wish to purchase. . . . .

Idea No. 2

. . . Pioneer Hi-Bred International, Inc., has done a very great amount of development work not only on early maturing corn varieties—but also on early maturity grain sorghum varieties. They are not only willing but anxious to have comparisons made in the Soviet Union between your own grain sorghum varieties and Pioneer brand varieties. . . .

Dr. [William L.] Brown[1] assured me Pioneer Hi-Bred International would be happy to furnish you samples of both their Pioneer brand hybrid corn and sorghum varieties without cost other than transportation from your Washington embassy to the Soviet Union . . . .

Pioneer Hi-Bred International, Inc., is making this same offer to Rumania, Hungary, Bulgaria, and Poland.

I wish to add that I am very excited about our new early maturing sorghums for several reasons.

First, in the United States grain sorghums are almost exclusively used in unirrigated land where the rainfall is less than half a meter.

Grain sorghum has an ability to wait longer for a rain before the plant is seriously damaged. A good deal of Bulgaria, Rumania, Hungary, and the Soviet Union and Poland has annual precipitation of less than five-hundred millimeters. So grain sorghums may well yield more grain than corn.

Corn is 10 percent more efficient if it is combined at "high" mois-ture—that is, when the grain contains 25 percent or 30 percent moisture. It can be ground and stored in trench silos without any loss—covered with a few inches of silage and polyethylene, and it will produce 10 percent more beef or milk or pork than when dry. It is widely used in the large western feed lots.

Where high moisture corn is 10 percent more efficient than dry corn, high moisture grain sorghum is 18 percent more efficient than dry grain sorghum. And it will combine perfectly at 25 percent moisture or even 28 or 30 percent; when the grain has been combined at 25 percent moisture, the

1. In 1971, vice president of Pioneer Hi-Bred International.

leaves and stalks—the grain sorghum stubble—is as green as grass in June. It can be chopped into very excellent grain sorghum stubble stover silage.

And you leave no trash on the ground. This makes stir plowing entirely unnecessary. All you need to do the next spring is to disc the ground, plant the sorghum, spray it with a herbicide—and harvest it. It is what is called "minimum tillage. . . ."

Idea No. 3

Irrigation—and especially the new "walk around" types of sprinkler irrigation systems.

It seems to me that the Ministry of Agriculture of not only the USSR but of Bulgaria, Rumania, Hungary, and Poland should send delegations to the USA next summer to study the very great present production of corn, sorghums, sugar beets, etc., on land that was not flat enough to irrigate by running the irrigation water between the rows.

Until about twenty years ago, running the irrigation water between the rows was the standard way to irrigate. Then came "pipe" irrigation where the pipes were moved by the use of hand labor and sprinklers were raised from the pipe that lay on the ground to a point a few feet higher than the crop to be irrigated.

But finally a method was perfected where an irrigation well was put down and a pipe on wheels was designed which would pivot on a circle automatically. They are very widely used—very effectively used in Nebraska, Kansas, Colorado, Oklahoma and Texas. They have proved to be of very great value. They require far less water per acre than the "between the row" type of irrigation—especially on sandy land.

Idea No. 4

At least one delegation—probably the one most interested in irrigation—should specialize on southwest Kansas—specifically the Garden City area. This area has made the greatest progress in agricultural technology in the United States in the last decade.

To illustrate, corn acreage in the western third of Kansas has expanded fivefold since 1959—and yields have increased from about 40 bushels per acre to 112 bushels per acre on nearly 200,000 hectares in 1971. This is the highest corn yield of any comparable area in the world.

This growth was brought [about] by increased irrigation, by improvements in planting and tillage methods, by high moisture harvest and storage and the feeding of high moisture grain to improve feed efficiency.

The area now fattens more than two million head of cattle annually—10% of the nation's cattle with another 20% in feed lots spreading northward into Nebraska and south into the high plains of Texas.

We encouraged the farmers in this development because we wished to sell them Pioneer brand hybrid seed corn—and we were successful. Now two-thirds of all the corn in the area is grown from Pioneer brand seed. Nearly all of it is planted with a new, improved soil opener which my son, David Garst, and our district sales manager at Garden City [Don Williams][2] invented, manufacture, and sell. This opener plants the seed in a V slice. I enclose a folder describing it.

With much of the seed, water is put in direct contact with the seed at planting time to speed germination and improve yields.

The opener also has opened the door to no-plow tillage—since the V slice like a pointed stick allows the opener to go into firm ground. Now much of this crop with maximum yields is grown with a minimum of effort and expense.

We take great pride in this development in western Kansas because we led it. We taught the farmers to irrigate their crops. We taught them how to irrigate previously unirrigatable land by showing them center-pivot self-propelled sprinkler irrigation systems which do not require leveling the land and which can work on sandy soils where water will sink into the ground before it runs down enough length of row. Incidentally, the first such system was patented in the United States in 1956.

We taught them how to fertilize their crops, to control the weeds and insects, and to more efficiently plant and havest their crops—and we did this with very little help from our country's agricultural schools.

This has been a commercial development—and we would be pleased to show your delegation the tools and technology that made it possible.

One part of this development—commercial cattle feeding—should be of extreme interest to the USSR. In scope and size there is much that could by adopted by your large state and collective farms. To illustrate, now 25 percent of all cattle in the United States are fattened in only four-hundred feed yards, 50 percent of all the cattle are fattened in only about two-thousand of them—and most if not all of this technology could be transferred to the USSR with very little cost.

Here again our leadership in teaching the use of urea as the principal source of protein has been the key. We first fed urea for all the protein in

2. Williams invented the planter-shoe, called "Accra-Plant," and later went into partnership with David Garst.

1947. Now in the United States, cattle feeders use over 600,000 tons principally in the fattening ration.

October 6, 1972

To Ms. Olga Iventyeva
Mr. Ivan Tikhomirov
Mr. Sergey Yefimov
Mr. Aleksandr A. Konygin[1]

It gives me pleasure to welcome you to the United States and to be permitted to show you about our agriculture. . . . We have entertained quite a few delegations. . . since 1955. It has given me a feeling of satisfaction to contribute some ideas that have been found helpful.

It has always seemed to me that with the world population growing as rapidly as it is the whole world must learn how to produce not only more food but better food—that is, more of the high type of protein for human consumption.

The world can produce enough starches in the form of wheat, rice, corn or grain sorghum, but the world needs more milk and cheese, more poultry and eggs, and more beef and pork.

. . . In the years 1930 through 1934, we raised more than 100 million acres of corn—with an average yield of 21.9 bushels per acre for an average total of 2,023,906,000 bushels.

In the last three years, we have averaged less than 60 million acres with an average yield more than 80 bushels per acre for a total of 4,924,823,000 bushels per year.

Only 60 percent as many acres—2.5 times as many bushels.

. . . As I have pointed out, I have seen quite a lot of Eastern Europe. I know that you do not have enough rainfall—not enough precipitation for maximum yields—in a great deal of the Soviet Union.

But that is true with large parts of the United States. As I will show

---

1. Olga Iventyeva: state inspector, Central Administration on purchase of grain and oil plants of USSR Purchasing Ministry.
Ivan Tikhomirov: deputy chief of Central Administration of Flour Milling and Grain Industry, USSR Purchasing Ministry.
Sergey Yefimov: chief of Central Administration on purchase of grain and oil plants, USSR Purchasing Ministry.
Aleksandr A. Konygin: agricultural counselor, Embassy of the USSR, Washington, D.C.

you on your short trip, we plant more acres of sorghum than we do of corn—unless we can irrigate the corn with ample water.

Grain sorghum will wait longer for a rain than corn will. In the western one half or even the western two-thirds of Kansas and Nebraska, for instance, where the precipitation is between 600 and 450 millimeters, farmers prefer to grow grain sorghum instead of corn in unirrigated land. But, they prefer corn to sorghum under irrigation. . . .

Furthermore, as you will be able to see, more and more farmers are not rotating crops. They take their best land and put it in continuous corn. And that practice lets them irrigate it with greater profit—because corn is the highest value crop.

Then they do not plow it—that is, they do not "stir" plow it—they just disc it and plant.

And, of course, I want to show you that the offal of the corn and sorghum plants make excellent cow feed (cattle feed) if they are accompanied with urea fed with a fast carbohydrate such as molasses or corn fed properly with minerals and vitamin A.

Please understand that it is a real pleasure to me to show you the progress that agriculture has made in the United States in the last forty years. Actually, most of our progress came after World War II. That was only twenty-six years ago.

. . .The average yield of corn in the United States in the five-year period of 1945 through 1949 was only 35.6 bushels per acre, and only up to 38.5 from 1950 to 1954.

And, corn yields doubled in the next fifteen years in the United States. Because I have been very active in the whole forty-two year period, I am bold enough to tell you how—the methods we used to increase yields to such a phenomenal extent.

I will list them for you in what I believe is their order of importance. First and by all odds the most important was the very greatly increased production of fertilizer. From a production of less than 2 million tons of nitrogen in 1952, we now have in the USA facilities that will let us fix 20 million tons. And as our production of nitrogen went up, we increased our production of acid phosphate P_2O_5—and our production of K_2O potash. . . .

The next most important item of increased yields is to control insects and fungi and weeds chemically so that rotations are not necessary—which lets us use the very best land for our most important crops.

We only have a few crops in a broad area of the USA that are profitable. Corn and grain sorghum are our most productive feed grain crops—

and soybeans are our most productive crop for the production of oil (from which we make our oleomargarine) and our high protein feed for our chickens and our swine.

Our food grains are largely wheat and rice. It will interest you to know that as our corn yields increased, we used fewer and fewer acres for the production of corn—and as the acreage of corn was reduced, the acreage of soybeans increased so that in a broad way the acreage of corn and soybeans combined has remained at about 100 million acres.

And we have found that an acre of corn for silage will produce about four times as much beef or milk as an acre of alfalfa or clover.

So, the amount of corn used as silage has gone up and up and up.

We have doubled our beef production since 1950 without increasing our acreage of hay, and I am confident that leguminous hays such as clover and alfalfa will gradually go down because it is cheaper to fix nitrogen mechanically than it is to raise it.

When Minister of Agriculture Matskevich first visited the USA in 1955, we were using a simple mixture of molasses and urea for all of the protein of our cattle—but we were among the very few people in the United States that were using it. I believe that in 1955, less than fifty-thousand tons of urea was being used as protein of ruminants in the USA.

In the summer of 1971, the U.S. Department of Agriculture estimated that between 550,000 and 650,000 tons of urea was being used as the protein of ruminants in the United States.

It just seemed to me that you should all know that I want to be as helpful as possible in every way I can while you are here in the United States on your short visit. And, it seemed to me that to write out my thoughts for you at the very start would be a good way. . . .

> Most sincerely,
> Roswell Garst

January 2, 1973

To Mr. Harrison E. Salisbury
New York, New York

Dear Harrison:

My nephew, John Chrystal, and I went to the Soviet Union, Rumania, and Hungary in November. . . .

When Khrushchev was demoted, I thought that I might not be welcome—and I think that might have been true before his death, but certainly not since.

In 1971, when Minister of Agriculture Matskevich visited the USA, both John Chrystal and I visited with him at length. We pointed out that in the USA almost no corn is planted in areas where the precipitation is less than twenty-four to twenty-six inches. With less precipitation than that grain sorghum is planted because it will wait longer for a rain without damage.

The only corn planted west of the 100° meridian is irrigated corn. The dry land is all planted to sorghum. The western two-thirds of Nebraska, Kansas, Oklahoma, and Texas are all from twenty-six inches to as low as sixteen or eighteen inches rainfall. They only raise winter wheat or grain sorghum unless they irrigate—then it is corn.

Matskevich accepted a present of twenty bags of grain sorghum seed. In October the Soviet Embassy asked if I could host a small delegation to study grain sorghum production.

I showed them grain sorghum production in Iowa, Nebraska, and Kansas, and they invited John Chrystal and me to come to the Soviet Union. They have ordered rather a substantial amount for more widespread testing in 1973, and I feel sure it will help them increase livestock production.

Then we went down to Rumania and Hungary. Both countries will, they have indicated, also buy enough grain sorghum seed for wide background.

While we were in Rumania, we went out on Saturday to Fundulea, their research station. They had about fifiteen or twenty of their top agricultural people and we spent several hours. It was, I believe [in Rumania], in the fall of 1956 that you first ran across my tracks and wrote an article for the *Times* about my having been a good diplomat by using a secret agent— No. 301—which a paragraph or two later you identified as Pioneer brand hybrid seed corn variety 301. One of the Rumanians at the table told the story. . . .

Sincerely,
Roswell Garst

April 10, 1973

From Dmitriy S. Polyansky
Minister of Agriculture
Moscow

Dear Mr. Garst,

. . . The nice meeting and discussion we had with you in Moscow are still fresh in my memory. I listened with great attention to your vivid description of sorghum and couldn't help believing in the great potential of this feeding crop. I hope that in future we shall have many more opportunities to discuss the sorghum problem with you and to take further steps for wider utilization of this crop to serve agriculture.

We are grateful to you for the delivered hybrid sorghum seeds. The seeds have arrived safely at our port and our experts have found them to be in a good condition. Broad comparative studies of these hybrids arranged by us in various parts of our country will enable us to determine which of them will turn out more suitable for the conditions of the Soviet Union. The results of these tests will make it possible to take the right decision at the end of this year concerning purchases of foundation lines of most suitable hybrids of grain sorghum as well as concerning purchases and sites of construction of sorghum seeds processing plants proposed by you.

I would like to thank you sincerely for the invitation to visit the USA. When possible I will take an opportunity to make such a visit.

With my best regards,
Dmitriy S. Polyansky

August 24, 1973

From Dmitriy S. Polyansky
Minister of Agriculture
Moscow

Dear Mr. Garst,

Thank you for your letter and kind words addressed to the people working in agriculture of the Soviet Union, who, as you said, have played great part in the progress of this most important branch of social production.

We have always attached and are attaching great importance to contacts and exchange of experience with foreign scientists, experts and farmers. This promotes the scientific and technical progress in agriculture and serves the cause of improving relations among countries. Therefore your invitation to send our delegations to the US is accepted with gratitude.

This year approximately in the second half of September the Ministry is planning to send to your farm a delegation of 4 specialists and an interpreter for 20 days. We shall be grateful if you acquaint them with problems of selection, cultivation and seed-growing of sorghum and corn, as well as with the experience of fertilizer application. It would be also useful for the delegation to acquaint itself with the work of some research institutes, experimental stations and other institutions, dealing with those problems. We shall additionally inform you about the composition of the delegation and the exact time of its arrival.

Next year in August or at any other time at your convenience the Ministry can send to your farm 3–4 specialists for the same period to study up-to-date agricultural technology used for corn and sorghum growing.

Please, let us know whether our proposals are acceptable to you.

As far as my trip to the US and my visit to your farm are concerned I shall be pleased to avail myself of your invitation as soon as such an opportunity comes.

Sincerely yours,
Dmitriy S. Polyansky

November 6, 1973

To Dmitriy S. Polyansky
Minister of Agriculture

Dear Minister Polyansky:

As I wired you and Export Khleb today, we are unable to completely fill your desire for twelve hundred tons of Pioneer brand hybrid grain sorghums, but we have done the best we can. While Garst & Thomas are furnishing the bulk of the seed from our Garden City plant, the balance will be furnished by Pioneer International from their plant at Plainview, Texas.

Your delegation headed by Mr. [Ivan] Khoroshilov[1] got to see both plants at harvest time. And your delegation had an opportunity to learn about the excessive demand for all kinds of equipment in the USA that now exists.

For instance, we visited with International Harvester, John Deere,

1. A distinguished Soviet geneticist.

and Massey Ferguson implement dealers both in Kansas and Iowa. They cannot promise deliveries on their combines nor even their most modern tractors in less than eight months in most cases.

What seems to have happened is that the world population has caught up with the world's food production ability. The whole world wants to improve the protein level of diets—and the industrial nations can afford to do so.

. . .We in the United States have one advantage over you of the Soviet Union—and you have a different advantage over us. We have a better climate for grain production. You, however, have more oil, gas, and urea.

It may well be that we should produce two or three million tons of corn, wheat, barley, grain sorghum and/or soybeans and trade them to the Soviet Union for an equal number of tons of urea on a year-to-year basis.

We are, as a nation, short of oil and gas. You, as a nation, are short of grain and with ample gas. If you use your natural gas to make dry urea, the same boats that hauled grain to the Soviet Union could haul nitrogen in the form of urea back to the USA.

I did not intend to write this kind of a letter—but I believe the suggestions I have made in it are logical suggestions, so I will send them on. . . .

Sincerely,
Roswell Garst

May 6, 1974

From Ivan Khoroshilov
Chief, Main Division of Cereals
Ministry of Agriculture
Moscow

Dear Mr. Garst,
I must advise you that early this year my health has gone bad. I spent about three months at a hospital and had to ask the permission to retire.

I would like to thank you very much for the attention given to me during my several trips to the United States of America, for your sincere striving to share with us your rich experience in corn, sorghum and animal production, for your kind assistance in acquainting us with your country's agriculture. Talks we had with you on these subjects were always pleasant and useful. My colleagues and I were greatly impressed by our latest trip during which you drove us 4 thou. km through states producing sorghum, corn and soybeans.

I hope that the relations of mutual understanding and cooperation established during the past years between you and the USSR Ministry of Agriculture will be further successfully developed. . . .

Sincerely yours
Ivan Khoroshilov

August 16, 1974

From Ivan Gavva
Agricultural Counselor
Embassy of the USSR
Washington

Gentlemen:

On behalf of the Ministry of Agriculture I extend to you a cordial invitation to visit the Soviet Union for ten days at the end of August 1974.[1] Our officials would like to discuss with you cooperation in the field of corn and sorghum breeding as well as the possibility of purchasing from you equipment for several seed plants. It would be very desirable if you could bring with you some papers concerning the subject (technical projects on the plants, a price list of equipment). All your expenses in the Soviet Union will be paid by the Ministry of Agriculture of the USSR. Enclosed are copies of the visa application form [for] your attention which we will need if you can make this visit. I am looking forward to developing future cooperation between our two Nations.

Very sincerely yours,
Ivan Gavva

---

1. The invitation was extended jointly to Garst and Chrystal.

November 21, 1974

To Mr. Boris V. Shapiro
Deputy Director of Department
V/O Traktorexport
Moscow

Dear Mr. Shapiro:

Today I am saying good-by to the Soviet delegation, headed by Mr.
V. Skliarenko, which has spent a good part of the last month in the U.S.
Midwest. They are intelligent, enjoyable, industrious people and we at
Garst and Thomas made every effort to show them the corn and grain
sorghum seed industry. . . .

The delegation and Garst and Thomas now have a complete mutual
understanding of what is desired for the grain sorghum seed plants and corn
seed plants. Due to the shortage of time. . .we cannot send a complete
offer home to Moscow with the delegation. We will forward to you in
approximately three weeks' time, three copies of a detailed offer of sale of
the seed plants and agricultural machines. This will include schematic
drawings, sectionalized drawings, and individual prices on the material to
be purchased in the United States. Also included will be the price of the
engineering drawings for the plants to be completed after a contract is
signed. These engineer drawings will be in the metric system and Russian.
The sale-offer schematic drawings will be as much as possible in the metric
system and Russian. . . .

Enclosed is a somewhat revised contract drawn by our attorney. I am
sure that there will need to be changes from this proposed contract, but I
am not sure what those final changes would need to be so I am only sending
it for further consideration and remarks from you. Any problems could be
worked out in a subsequent meeting between Garst and Thomas and you
after you see our offer in the next three weeks. You will note that Garst and
Thomas will provide much more "on site" consultation than first thought.

We feel perfectly confident that we can provide the USSR with efficient seed plants at a reasonable cost. You understand that the U.S. economy is in a state of flux so that delivery will vary on various items. While we
will make every effort to have the material available for construction so that
the plants could handle the 1975 harvest—certainty of that construction
completion is most doubtful. Garst and Thomas started construction of a
new corn plant in early winter of 1973. It was built by workers with much
experience with the seed industry. That plant was barely usable for the
harvest of 1974.

Garst and Thomas sells and produces its seeds in the area of the United States most comparable to the Soviet Union in climate. We sell the highest percentage of corn and sorghum seed sold in the area. We believe, and I assume the Soviet delegation observed, that we take greater care to see that we produce higher quality seed than any of our competitors. We have 40 years experience in producing this high-quality seed for the western Corn Belt of the United States. . . .

Sincerely,
Roswell Garst
John Chrystal

June 6, 1977

From Alexander Tulupnikov

Dear Mr. Garst:

My son Andrei, bearer of this letter, has a 3 month tour in the USA for studying of agriculture, on the program of the cooperation between the USA and USSR.

Andrei works in the Institute of the USA and Canada Study (with Mr. [Victor] Lischenko). His program includes studying of Iowa's farm for acquaintance with technology of corn, beef and dairy production. I should like this farm to be Garst & Thomas Co. I appreciate your ideas in respect to utilizing urea in beef industry, and also to growing corn as monoculture, without rotation with grasses. About your practice and economical substantiation I talk in my works, as I see, like you, in it important condition of increasing production milk, meat and eggs in my country. I should like Andrei to better understand and learn economical substantiations of your farm's business.

I should be thankful to you if you find possibility to spend some time [with] Andrei and help him to solve his problem.[1]

Sincerely Yours
A[lexander] Tulupnikov

1. Holograph letter.

Appendix

AGRICULTURAL OPPORTUNITIES OF THE USSR AND ALSO THE
AGRICULTURAL RESPONSIBILITIES OF THE USSR

By Roswell Garst
A Farmer from Iowa in the United States

As an American farmer who is familiar with the progress being made
in the agriculture of not only the Soviet Union but the other countries of
Eastern Europe such as Hungary and Rumania, I welcome this opportunity
to compare the agriculture of the USA with that of Eastern Europe—and to
suggest what I believe may be helpful.

But at the very start, I want to point out the very great differences in
the background of your country and my country.

Until 1917, the government of Russia was despotism. Your country
had fine schools for the aristocracy—where a few people could get a very
advanced education.

But the great bulk of the citizens were illiterate.

That was forty-seven years ago! So first I want to remind you that
opportunitywise you are a very *young* country.

It must have taken ten or twelve years to organize a government and
to educate the teachers who were going to educate the mass of your citizens
and to build the schoolhouses.

Without general education, no nation can make progress—no nation
has opportunity!

So let's cut down the time of opportunity from forty-seven years to
thirty-five years.

Then the fighting of a war of survival—and the recovery from the
effects of that war—took another ten or twelve years away. So you have—
as a nation—had somewhere between twenty and twenty-five years of
opportunity.

Similarly, Hungary and Rumania were feudal governments with mass
illiteracy until after World War II, so a somewhat similar handicap existed
in those countries as well.

Now in the United States, we have had a democratic government for
nearly two-hundred years. We have had free and fine schools for several
generations. We are a hybrid people—a cross of the people who were brave

enough to leave for a new land. We have had a high rate of literacy for generations.

Furthermore, we have not had a war on our soil since the Civil War ninety years ago. You, on the other hand, have had two devastating wars in the last fifty years.

So——

When I first came to the Soviet Union in 1955—and then went through Hungary and Rumania and Czechoslovakia on my way home—our agriculture in the United States was far, far ahead of the agriculture of Eastern Europe, not because we are brilliant people and you are stupid people but because of the circumstances I have just outlined. We have had better circumstances.

As a matter of fact, the agricultural progress in the United States made very, very, slow progress until after 1940.

The yields of our crops—of all crops—gained very slowly from 1890 to 1940. The yield of corn, for instance, stayed steady at twenty-six bushels per acre from 1890 to 1930.

The yield of wheat stayed at about sixteen bushels per acre—and the yield of cotton at about one-third (1/3) bale per acre.

And it required almost the same number of hours to raise a ton of grain in 1930 as it had required in 1910.

It was from 1940 on that the agricultural productiveness of the United States exploded and the labor per ton of grain went *down—down—down!*

And during the first fifteen years of that period, from 1940 to 1955, the people of Eastern Europe—not only of the Soviet Union—but of Hungary, Rumania, Poland, Bulgaria, Yugoslavia, and Czechoslovakia as well—were fighting a war of survival and recovering from the destruction of that war.

When the war was finished you had to rebuild your transportation systems, your factories, your electric systems, your homes. And your war casualties were greater than those of all other countries combined—so you had to do it with a shortage of men who had been killed.

These things I know!

During the war and postwar period the United States had to raise a large army and get it to Europe; we had to build ships and lots of armaments, not only for ourselves but for our allies, and we had to produce food not only for ourselves but for our allies.

Where Eastern Europe could not produce normal quantities of food, the USA had to produce *much more food than ever before*—and had to do it with *much less labor!*

So——

In 1955, we farmers of the United States were far more productive than the farmers of Eastern Europe—not, I wish to repeat, because we are brilliant and you are stupid and lazy but because of a difference in circumstances.

Because the people I met in the agricultural institutes in Eastern Europe in 1955 were fine, intelligent, industrious people and because the people I met in the state and collective farms were industrious and intelligent and were ambitious to make the greatest progress possible, I came to the conclusion that while the agriculture of Eastern Europe was, in 1955, where the agriculture of the United States had been in 1925—that is, thirty years behind us—you should be able to shorten that time to one-third or one-half the time it had taken us.

When my nephew, John Chrystal, and I visited the Soviet Union in the spring of 1963, I stated that I thought you were eight years behind; you will not catch us in eight years because we are still making great progress. But you will, I predict, have doubled your crop yields—and cut your labor requirements very greatly.

With the foregoing preface let us study what the United States has done, how we have done it, and what Russia and all of Eastern Europe has done and can and will do—

The first really big step in agricultural productiveness was the change from varieties of corn to hybrids of corn.

Hybrid seed corn is universally used in the United States—not only where corn is planted for grain but also where the corn is used for silage.

It took fifteen years—from 1930 to 1945—for the United States to change from very little hybrid use to almost complete hybrid use.

It is interesting that it took only five years—from 1956 to 1960—for the Soviet Union to go from very little hybrid to almost complete use of hybrids for grain, although the use of hybrids for silage is not complete yet and *should* be without further delay.

Hybrid seed corns have gained rapidly in the other countries of Eastern Europe and should be universally used as soon as possible.

Hybrid corn stands up better so it permits better use of mechanical equipment and on the average it yields 30 percent more than the varieties, and its use is extremely profitable. It should be universally used everywhere and immediately.

Because the United States raises about one-half of the corn in the world, and because we first mechanized corn growing, and because we are a highly mechanized nation, our most modern corn growing machinery is more advanced, easier to handle, and less complicated than the machinery of Eastern Europe.

I would think it advisable for every country in every part of the world to buy and use some of the newest and most advanced American farm machinery each year.

The improvements in American farm machinery design have been astonishing—and the improvements seem to be greater each year than they were the year before.

Consider this fact!

When Mrs. Garst and I were married forty-two years ago, it used to take thirty minutes of man time to produce a bushel of corn. Now the average American farmer produces a bushel of corn in four minutes—and the best farmers produce it in two minutes. There are forty bushels of corn in a metric ton, so a good farmer in the USA can produce and harvest a ton of corn in two hours.

Those are two big steps we have taken, that is, hybrid corn and mechanization.

In each case, the agriculture of Eastern Europe has taken those same two steps. Eastern Europe is using a high percentage of hybrid seed corn, but not all hybrid corn, and Eastern Europe is mechanizing, but not as well as is possible.

A third major development in the United States was the extra ordinary increase in the use of fertilizer. It has greatly increased the yields of both the food grains (wheat and rice) and the feed grains (corn, grain, sorghum, barley, and oats) and the fiber such as cotton. To keep it simple let us study corn yields—because the yield of all farm crops has gone up about the same. . . .

For forty years, 1890 to 1930, the yield stood at about twenty-six bushels per acre.

Then, for fifteen years it went up each year as the percentage of hybrid seed went up. Hybrid seed use pushed yields from twenty-six bushels to thirty-eight bushels, where they stayed steady for about ten years.

By the early 1950s we were short of food in the United States. We had been producing only enough fertilizer to demonstrate what an increase we could get.

But in 1952, and from then on for several years, the United States government stimulated the building of fertilizer plants by favorable tax treatment so we started building fertilizer plants as rapidly as possible.

It took several years to build the large, efficient fertilizer plants so it was not until the late 1950s that the increased production occurred.

Just look what has happened to yields:

Corn yields which used to average twenty-six bushels per acre before 1930 hit sixty-seven bushels per acre in 1963.

The United States is continuing to build fertilizer plants at a rapid rate and will continue to improve yields.

Not only is the United States using more fertilizers—and better balanced fertilizer—we are using better insecticides and better herbicides.

Now the announced plans are for the Soviet Union to do exactly those same things—to build ample fertilizer production units and at the same time to build chemical facilities for insecticides and herbicides and . . . additional facilities for the enlarged production of vitamins and antibiotics particularly for the feeding of the swine and poultry in which vitamins and antibiotics are so important.

As I pointed out about the farm machinery so I also wish to point out about chemical plants.

The United States has very advanced engineering in the building of chemical plants. Our fertilizers are beautifully pelleted—and therefore extremely easy to apply.

This was not true as little as fifteen years ago. We had fertilizer that was either too dusty or caked into big chunks and was almost impossible to spread.

Now *all* American fertilizer is granular, about the size of rice or grain sorghum, and it can be easily spread.

Unless fertilizer is of high quality and easy to apply, there will be resistance to its use by the farm worker. Also, we in the United States have very simple machinery for the spreading of high quality fertilizer. No machinery ever works well with low quality fertilizer.

I know that some of the fertilizer plants of Western Europe make pelleted fertilizer—some do not—and I wish to emphasize the absolute necessity of high quality manufacturing so the product can be easily used.

I am certain that the heavy investment planned in chemical plants for the Soviet Union is completely wise and [they] will pay big dividends in increased food production as soon as they are completed.

But one more thing is, in my opinion, utterly necessary, and that is better roads!

I do not mean the highways that lead between cities. You have fine highways! I mean what we, in the United States, call "farm to market roads." The United States has more and better farm to market roads than any other area of its size in the world.

We have done it gradually over the years.

And we have much the best designed road building machinery in the world for the building of these "farm to market" roads.

With a set of American road building machinery, five or six men can build a kilometer of a road per day without difficulty. They grade the road up so the top of the road is about one-third of a meter above the level of the countryside—with a ditch on each side of the center about 1 meter below the level of the country.

The higher center part of the road is about ten or twelve meters wide and slightly higher in the center than on the sides so the snow blows off— and the rain runs off.

I have repeatedly recommended that the Soviet Union and the other countries of Eastern Europe buy several sets of American machinery to see for themselves in their own countries what can be accomplished with a minimum expenditure of money if the right tools are available.

Good farm to market roads are essential for an advanced agricultural economy. Fertilizer and tractor fuel and other supplies must get to the farm in any weather. And the produce of the farm must get to the center of population.

Good roads are not an expense! They save time! They save trucks! They save tires! Good roads are a necessity—*not a luxury!*

Now let us list them again:

1. Hybrid seed corn—The United States uses 100 percent, and Eastern Europe approaches that.

2. Mechanization—The United States is the most highly mechanized, but Eastern Europe is progressing rapidly.

3. Fertilizer—The United States built [fertilizer plants] in the 1950s as Eastern Europe is building in the 1960s.

4. Roads—The United States is far ahead, and Eastern Europe needs to get started immediately and make steady progress from now on.

In the nine years from 1955 to 1964—the nine years since I first visited Eastern Europe till now—the progress has been excellent. I feel sure that the progress will be even more rapid from now on. I am confident because I have found the people of the Soviet Union to be hardworking, intelligent, and friendly people, who want their country to prosper, want their living conditions to improve, and realize that this can happen only by the adoption of the most modern methods.

I have several other agricultural suggestions that can be of importance to the agricultural production.

All of Eastern Europe is handicapped by a lack of high protein supplement for livestock and poultry production.

In the United States we have very large areas where the weather and the soil are ideal for the production of soybeans. We press out the oil to use for margerine—and we have a 42 percent protein soybean meal we can use for the protein supplement of our poultry, swine, and cattle.

Unfortunately, Eastern Europe is not so well adapted to soybeans. So sunflowers are raised for your oil and meal. While sunflowers yield well— produce high quality oil and meal—they are hard to raise, hard to mechanize!

So Eastern Europe does not have an almost unlimited supply of high protein supplement such as our soybean meal.

There is however, a way that Eastern Europe could greatly improve the feeding situation.

Urea, a nitrogen product, which will be manufactured in very large quantities, not only in the USSR, but also in Rumania and Hungary, can be used as all of the protein of cattle and sheep.

The amount of urea required is very small! Where cotton seed meal or soybean meal contain about 42 percent protein, urea contains 262 percent protein equivalent.

Urea is $6\frac{1}{4}$ times as high in protein equivalent so you need to feed only $\frac{1}{8}$ as much urea as you would feed of cotton seed or soybean meal.

Urea cannot be fed to poultry or swine; it is only good for ruminants— the cattle, sheep, goats, or water buffalo of Eastern Europe, or the camels.

If urea were substituted for the high protein supplement of all ruminants, however, and the by-product meals those ruminants had formerly eaten were fed to the poultry and swine, it would permit the expansion of not only the ruminants—the cattle, sheep, and goats—but the poultry and swine as well.

The agricultural technicians should insist on the rapid use of urea as the high protein of all the ruminants. That way all of the sunflower meal, flaxseed meal, cottonseed meal, and fishmeal could be used for the poultry and swine.

In some things, the agriculture of the Soviet Union is ahead of the agriculture of the United States. An example is the artificial insemination of sheep, which is so widely practiced in the Soviet Union—and never in the United States.

I have never seen such uniform flocks of sheep as I have seen in the North Caucasus area of the Soviet Union.

As Mr. Shevchenko wrote to me recently, if two men each have an apple and they exchange apples they still each have one apple.

But if two men exchange ideas, each man ends up with two ideas.

It is a fine thought and I agree with it. It has given me great pleasure to have had many delegations from not only the Soviet Union but Hungary, Rumania, and Bulgaria as well, visit our farm—and for me and Mrs. Garst to have visited and seen so much in the countries mentioned.

It is fortunate for the Soviet Union to have a farmer in the person of Chairman Khrushchev at the head of the government. Both the chairman and Madame Khrushcheva are very well informed about agricultural matters and aggressively interested in seeing the greatest kind of progress made.

Since 1950 the world has been spending about $100 billion a year, preparing for a war nobody wants, a war that nobody expects, a war that no one could survive. Such spending on armaments was global insanity. It has been unbelievably stupid. It now looks as though we may be getting enough of such a waste of money and energy.

The United States and the Soviet Union have both announced reductions in the expenditure for armaments for the coming year.

The reductions in the expenditures are too small, but at least there is some reduction and discussions are going on that may bring further reductions.

There is no doubt that the people of both countries hope for further and major reductions.

What progress we could make if, instead of armaments we could build more homes, more roads, more and better schools, factories, hospitals, and the things that let people live better.

I have never met an American—nor a citizen of the Soviet Union—who wanted war, nor citizens of any other country who want wars.

What all people want is for the disarmament conference to be successful so that we can build things that will bring us a better world in which to live.

The world now has more than 3 billion people.

By the year 2000 it will have more than 6 billion people.

A great many of the 3 billion now do not get a good diet.

The knowledge on how to produce more and better food—and produce it with much less labor—now exists.

The industrial nations of North America and Europe not only can—*but must*—help the less fortunate nations of Central and South America, Africa, and of the Mideast and Asia.

The United States and the Soviet Union and Western Europe must not only build chemcial plants for themselves—they must build them in the less fortunate coutries—and not after a while but in the nearest future.

The world can no longer afford to spend $100 billion a year on armaments the world dares not use.

Once that is understood and we quit that foolish expenditure, the world can feed and house its citizens and educate them.

So it is my great hope that at the February meeting of [the] Central Committee not only will the intensification of agriculture be discussed but disarmament progress will also be emphasized.

My hope is that great progress can be made in both fields—in agriculture and disarmament.

Roswell Garst
[January 1964][1]

1. Published 13 February 1964 in *Pravda* and *Izvestiia*.

Index

Lightning Source UK Ltd.
Milton Keynes UK
UKHW040746290919
350631UK00003B/205/P